THE USDA 2007 FARM BILL PROPOSAL

JASPER WOMACH,
GEOFFREY S. BECKER
AND RALPH M. CHITE

Nova Science Publishers, Inc.
New York

Copyright © 2008 by Nova Science Publishers, Inc.

All rights reserved. No part of this book may be reproduced, stored in a retrieval system or transmitted in any form or by any means: electronic, electrostatic, magnetic, tape, mechanical photocopying, recording or otherwise without the written permission of the Publisher.

For permission to use material from this book please contact us:
Telephone 631-231-7269; Fax 631-231-8175
Web Site: http://www.novapublishers.com

NOTICE TO THE READER

The Publisher has taken reasonable care in the preparation of this book, but makes no expressed or implied warranty of any kind and assumes no responsibility for any errors or omissions. No liability is assumed for incidental or consequential damages in connection with or arising out of information contained in this book. The Publisher shall not be liable for any special, consequential, or exemplary damages resulting, in whole or in part, from the readers' use of, or reliance upon, this material.

Independent verification should be sought for any data, advice or recommendations contained in this book. In addition, no responsibility is assumed by the publisher for any injury and/or damage to persons or property arising from any methods, products, instructions, ideas or otherwise contained in this publication.

This publication is designed to provide accurate and authoritative information with regard to the subject matter covered herein. It is sold with the clear understanding that the Publisher is not engaged in rendering legal or any other professional services. If legal or any other expert assistance is required, the services of a competent person should be sought. FROM A DECLARATION OF PARTICIPANTS JOINTLY ADOPTED BY A COMMITTEE OF THE AMERICAN BAR ASSOCIATION AND A COMMITTEE OF PUBLISHERS.

LIBRARY OF CONGRESS CATALOGING-IN-PUBLICATION DATA

Womach, Jasper.
   The USDA 2007 farm bill proposal / Jasper Womach, Geoffrey S. Becker and Ralph M. Chite (authors).
      p. cm.
   ISBN 978-1-60456-813-4 (softcover)
   1. Agricultural laws and legislation--United States. 2. United States. Dept. of Agriculture. 3. Agriculture and state--United States. 4. Agricultural conservation--United States. 5. Agriculture and energy--United States. 6. Rural development--United States.. I. Becker, Geoffrey S. II. Chite, Ralph M. III. Title.
   KF1682.W64 2008
   343.73'076--dc22
                                    2008027827

Published by Nova Science Publishers, Inc. ✦
New York

THE USDA 2007 FARM BILL PROPOSAL

CONTENTS

Preface		vii
Chapter 1	Title I: Commodity Programs	1
Chapter 2	Title II: Conservation	15
Chapter 3	Title III: Trade	25
Chapter 4	Title IV: Nutrition	35
Chapter 5	Title V: Credit	57
Chapter 6	Title VI: Rural Development	59
Chapter 7	Title VII: Research	65
Chapter 8	Title VIII: Forestry	71
Chapter 9	Title IX: Energy	75
Chapter 10	Title X: Miscellaneous	83
Appendix. Administration's Cost Estimate		87
References		91
Index		93

PREFACE

On January 31, 2007, the Secretary of Agriculture publicly released a set of recommendations for a 2007 farm bill. The proposal is comprehensive and follows largely the outline of the current 2002 farm bill, which expires this year. It includes proposals regarding commodity support, conservation, trade, nutrition and domestic food assistance, farm credit, rural development, agricultural research, forestry, energy, and such miscellaneous items as crop insurance, organic programs, and Section 32 purchases of fruits and vegetables.

The Administration delivered its report to Congress, not as a bill, but as a possible focus for debate and a foundation for developing legislation. CRS has received many questions about the content of and potential issues related to the Administration proposal. Given the early stage of the debate, this book poses some questions that may contribute to a better understanding of the proposal.

This book contains a brief description of current policy on each topic, a short explanation of the Administration's proposals, and then questions of a policy, program, and/or budgetary nature. In some cases proposals are repeated in more than one title, and where this happens the questions are duplicated.

Secretary of Agriculture Mike Johanns, at a public meeting on January 31, 2007, described the Administration's 65 recommendations for a new farm bill. These recommendations were published by the U.S. Department of Agriculture (USDA) in a report subsequently transmitted to Congress titled *2007 Farm Bill Proposals*. The report and related USDA materials are available on the Department's website at [http://www.usda.gov/wps/portal/!ut/p/_s.7_0_A/7_0_1OB?navid=FARM_BILL_ FORUMS]. In the following pages, Congressional Research Service analysts summarize current policy and programs, describe the USDA recommendations, and pose questions related to the policy, program, and/or budgetary impact of the recommendations. The USDA report presumes a new five-year farm bill covering the 2008-2012 time frame. However, consistent with U.S. government annual baseline budgeting, the spending authority and spending outlay estimates of the Administration's farm bill are projected for 10 years.

In: The USDA 2007 Farm Bill Proposal
Editors: J. Womach et al.
ISBN: 978-1-60456-813-4
© 2008 Nova Science Publishers, Inc.

Chapter 1

TITLE I: COMMODITY PROGRAMS[*]

The 2002 farm bill mandated support for a group of commodities (grains, oilseeds, cotton, sugar, and milk) that long have received support, and it added six more commodities (dry peas, lentils, small chick peas, wool, mohair, and honey) to the list. The Commodity Credit Corporation (CCC) pays the costs of commodity support under a $30 billion line of credit from the U.S. Treasury. Congress annually appropriates funds to the CCC that are used to pay down its loans from the Treasury. Commodity support expenditures are estimated to have averaged about $12.6 billion per year over the six-year life of the current farm bill (FY2002-FY2007). The Congressional Budget Office (CBO) forecasts that future spending for commodity support under current law would amount to about $7 billion per year over the next five years (FY2008-FY2012). The decline from past spending levels is due to current and anticipated high market prices for supported crops well into the future.

USDA's proposed 2007 farm bill outlines modifications to the commodity programs that are claimed to save $4.494 billion from the Office of Management and Budget (OMB) 10-year current services baseline of $74.566 billion for commodity support. In addition to the financial savings, the modifications to current law, according to the Secretary of Agriculture, are designed to make the programs "more market-oriented, more predictable, less market distorting and better able to withstand challenge" in the World Trade Organization.

---

[*] Excerpted from CRS Report RL33916, dated March 12, 2007.

MARKETING ASSISTANCE LOANS

Current law specifies support prices for 25 commodities, including corn and other feed grains, wheat, cotton, rice, soybeans and other oilseeds, peanuts, dry peas, lentils, small chickpeas, wool, mohair, honey, and milk.

The Secretary asserts that crop loan rates (one of the support price mechanisms) are set at such high levels that they encourage overproduction and cause lower market prices. In contrast, according to the Secretary, the USDA proposal would minimize market distortions and encourage farmers to plant crops based on market prices instead of subsidy prices. *Loan rates for each commodity would be set at the lesser of (a) 85% of the five-year Olympic average of market prices (i.e., the average of the last five years excluding the high and low year), or (b) the loan rates specified in the House-passed version of the 2002 farm bill (which are lower than current law for feed grains, wheat, cotton, oilseeds, and peanuts).* This proposal is claimed by the Administration to save $4.5 billion over 10 years ($450 million per year) compared to baseline spending of $8.807 billion ($880.7 million per year) on marketing assistance loan program operations.

How much higher or lower would federal outlays have been if the proposal had been in place over the past five crop years?

Over the past five years, which commodities have experienced season average market prices below the levels proposed for the new loan rate formula?

How much of a decrease in savings would result if the loan rates specified in current law were used in the new formula instead of loan rates in the 2002 House-passed farm bill?

Has the idea of replacing nonrecourse loans with recourse loans been considered as a mechanism to eliminate commodity certificate gains or the forfeiture of commodities to the government? How much has been paid to farmers in the form of commodity certificate gains over the past five years? Are certificates used in order to circumvent payment limits?

POSTED COUNTY PRICES

Under current law, farmers are allowed to sign up for loan deficiency payments (LDPs) to secure the benefits of the marketing assistance loan program instead of taking out nonrecourse loans. Farmers can capture LDP gains when posted county prices (PCPs), which serve as proxies for county market prices, are lower than loan rates. This opportunity is available daily. The Secretary asserts that there are substantial difficulties and inequities in calculating daily PCPs. Additionally, short-term market price declines create windfall opportunities for farmers that are costly and inconsistent with the fundamental income support objectives of the marketing loan program.

The proposed change would replace daily PCPs with monthly PCPs. Additionally, farmers would receive LDPs based on the monthly PCPs in effect on the days producers lose beneficial interest in the commodities.

Currently, farmers can collect LDPs and hold onto commodities and market them at a later time when market prices are higher than the loan rates. The USDA proposal is claimed to save $250 million over 10 years ($25 million per year) compared to an unspecified OMB baseline spending level.

Over the six-year life of the current farm bill, LDPs are estimated by USDA to cost an annual average of about $2.547 billion. How much would the new proposal have saved had it been in place in the current farm bill?

DIRECT PAYMENTS

Direct payments were enacted in the 2002 farm bill as a replacement for production flexibility contract payments, which were first enacted in the 1996 farm bill. Direct payments are made on land with a history of production (called base acres) of feed grains (largely corn), wheat, upland cotton, rice, oilseeds (largely soybeans), and peanuts. The payment rate for each commodity is specified in the law. The annual payment to a farm is its commodity base acres times program yield times the payment rate. The direct payment program is designed to cost about $5 billion per year. The payment is made for each eligible farm based on historic production and yield, and not on actual production or market prices. Hence, it was envisioned as not trade-distorting and not subject to WTO spending limits on subsidies.

The Secretary has proposed to continue the direct payments and, except for upland cotton, to keep the current payment rates in place from 2008 through 2009, and then increase them by about 7% in 2010 for the three-year period through 2012. Uniquely, the upland cotton direct payment rate would immediately increase about 7% and remain at the higher level.

Is the increase in direct payments for program commodities a mechanism to maintain payments to farmers that otherwise would decline under conditions of high market prices and reduced marketing loan program payments? If market prices do remain high, what would be the economic justification for higher direct payments? Would these higher payments be capitalized into higher land prices and rents?

A high proportion of the increase in direct payments would be for one commodity, upland cotton. USDA's recommendations report states that "[t]he combination of increases in upland cotton yields per acre and declining U.S. upland cotton textile production is expected to limit price gains and result in substantial cotton program expenditures, compared to other commodities." Why would a declining domestic textile industry have a depressing impact on prices given that in the past, increased exports have offset any decline in domestic use? Are the loan rate and target price for cotton substantially out of line with markets, and the cost of production? Are marketing loan and counter-cyclical program costs likely to be high in the future compared to other commodities?

Over the past five years, crop disaster payments amounted to about $1.3 billion per year, and no payments yet have been made on disaster losses for 2005 and Would farmers be better served if the $5.5 billion proposed increase in direct payments were instead used for crop disaster assistance (or possibly through crop insurance) over the coming 10 years?

DIRECT PAYMENTS FOR BEGINNING FARMERS

When it comes to eligibility and payment rates for direct payments, current law makes no distinction among farmers with regard to age, longevity as farm operators, or ownership status. The eligibility requirement is that a person must be actively engaged in farming on an operation that has program commodity base acres. The sharing of direct payments between tenants and landlords is a matter agreed to among the parties in a manner consistent with local custom. *The Secretary's farm bill proposal would give beginning farmers direct payments at a rate that is 20% higher than for other farmers for the first five years.* The expected cost is $250 million over 10 years.

How does the USDA intend to define a beginning farmer?
How many beginning farmers are expected to benefit from this program?

Is it likely that sellers of farmland will raise the asking price or the beginning farmers will raise the offer price to acquire cropland by the amount of the increased direct payment? In other words, could the higher direct payment be bid into higher cropland prices and higher rental rates, as has been the case with regular direct payments?

REVENUE-BASED COUNTER-CYCLICAL PAYMENTS

Counter-cyclical payments were adopted in the 2002 farm bill as a way of providing certainty and stability to *ad hoc* emergency market loss payments enacted in years following low market prices. The five-year (FY2003-FY2007) average annual cost of counter-cyclical payments is estimated at about $2.5 billion. The payments are made when the season average farm market price of a program crop are below the effective target price. The counter-cyclical payments, like direct payments, are paid on base acres without regard to what or how much of any crop is grown on the base acres.

The Secretary has proposed that counter-cyclical payments be triggered by a shortfall in national crop revenue rather than a shortfall in the national average price. This change would bring crop yields and production into the equation. There have been years when prices were high but yields were low, so farmers were in need of support but the program made no payments. In contrast, there have been years when the price was low but yields were high so payments were made even though farmers did not need the support. This change is estimated to generate savings of $3.7 billion under the OMB 10-year current services baseline of $11.245 billion.

Historically, commodity support programs have been designed to absorb risks on the price side of the farm income equation, while crop insurance and disaster assistance have addressed yield risks. Does the idea of revenue-based counter-cyclical payments integrate the two sides of risk management in a way that could serve as a model to integrate commodity support with disaster assistance and crop insurance? What are the risks and benefits of integration?

What are the fundamental economic, business, trade agreement, or political barriers to creating a national counter-cyclical revenue program to replace commodity support programs, crop insurance, and disaster assistance for at least the subsidized crops?

With few exceptions over the past 20 years, Congress has provided disaster assistance to the nation's farmers and ranchers whenever weather-related losses have been substantial. Some would argue that the Administration's farm bill proposal does not appear to offer what the Secretary of Agriculture might call an "equitable, predictable" alternative for any farmers not producing program crops. Could the revenue-based counter-cyclical program contained in the farm bill proposal for program crops serve as a starting point for a similar program for unsupported crops?

VALUE AND ELIGIBILITY LIMITS

In 1970, Congress first enacted annual limits on commodity support program payments. Currently, there is a per person limit of $40,000 on direct payments, $65,000 on counter-cyclical payments, and $75,000 on marketing loan gains/loan deficiency payments. The combined limit of $180,000 can be doubled to $360,000 under the spouse allowance or the three-entity allowance (a person can receive payments on three farms but at half the value on the second two). There is no limit on commodity certificate gains or marketing loan gains from the forfeiture of collateral under the nonrecourse loan program. Also, there is an adjusted gross income (AGI) eligibility cap of $2.5 million.

In practice, according to the 2002 farm bill-authorized Commission on the Application of Payment Limitations for Agriculture, most farmers pushing up against the limits have devised ways to avoid the limits. In general, cotton and rice farmers (because of the higher per acre value of their crops and their large size) feel threatened by payment limits, in contrast to corn, soybean, and wheat farms, introducing a regional factor (north vs. south) into the debate. For some proponents of lower and more effective payment limits, it is an issue of equity and a concern that large payments accelerate the consolidation of farms into ever larger units. Opponents argue that the payments are fundamental to the safety net for agriculture and that large efficient farms are equally subject to risks as smaller farms.

The Secretary has proposed eight changes that would make it difficult to evade a $360,000 per individual payment limit and would exclude anyone with more than $200,000 in adjusted gross income from eligibility for commodity program payments. The proposal is expected to save $1.5 billion under the OMB current services 10-year baseline of about $75 billion for commodity support.

The projected savings of $1.5 billion from tightening the eligibility and payment limits appears small compared to the roughly $75 billion 10-year baseline. How many farms and how many individuals are expected to be affected?

How much of the savings would come from reduced cotton and sugar payments, the commodities likely to be most impacted?

A large proportion of commercial farms are managed by operators who own some land, but not even a majority of the acres they farm. In crop share arrangements, absentee landlords are receiving commodity program payments. To what extent would absentee landlords be impacted by the new $200,000 AGI limit? Could this lead to changes in tenant-landlord lease contracts toward cash rent? Would a further shift to cash rent be good for U.S. agriculture?

SECTION 1031 EXCHANGES

Section 1031 is a feature of the federal tax code that allows a seller of income-producing property to acquire like-kind property and treat the transaction as a tax deferred exchange. The capital gains on the disposed property are transferred to the basis of the acquired property, and taxes on the gains are deferred until the acquired property is sold at a later date. This feature of the tax code is well known and used by owners of rental housing, but the rule applies to all income producing property, including farmland.

The Secretary's proposal would prohibit commodity subsidy benefits to any farm acquired through a 1031 exchange. Justification is based on the argument that tax-deferred farmland exchanges are contributing to the escalation of farmland prices, making it difficult for new entrants to purchase land and small farms to expand.

Do farmers ever use the 1031 exchange to geographically consolidate their holdings for efficiency purposes? If so, has the Administration considered exempting these farmers from the proposed new policy?

Might there be cases where beginning farmers would use the 1031 exchange to acquire farmland? Should an exemption be made to the proposed prohibition in cases of beginning farmers?

Can the argument that Section 1031 creates economic distortions, contributing to increased farmland prices, be applied to other parts of the economy that utilize that feature of the tax code?

Several studies, including work done in the USDA, conclude that commodity subsidies are substantially capitalized into land prices and higher rental rates. What has a greater impact on farmland price escalation, commodity subsidies or 1031 exchanges? Is redesigning commodity programs to eliminate their effect on land prices something that should be considered, or not?

Have the Secretary of the Treasury and the Ways and Means Committee been consulted about changing the tax code to eliminate 1031 exchanges in general or for farmland in particular?

DAIRY COUNTER-CYCLICAL PAYMENTS AND PRICE SUPPORT

The federal government long has mandated that the farm price of milk be supported. The 2002 farm bill continued the Dairy Price Support Program at the then-current support price of $9.90 per hundredweight (cwt.) of farm milk. Support is achieved through a standing offer to purchase cheese, butter, and nonfat dry milk at prices equivalent to the milk support price. *The Administration proposal recommends continuation of the program at the current support level of $9.90 per cwt.*

The 2002 farm bill authorized a new counter-cyclical dairy payment program, called the Milk Income Loss Contract (MILC) program. Under the MILC program, dairy farmers nationwide are paid whenever the minimum monthly market price for farm milk used for fluid consumption in Boston falls below $16.94/cwt. As amended by the FY2006 Budget Reconciliation Act, farmers receive 34% of the difference between the $16.94 target price and the lower market price on up to 2.4 million lbs. of annual production. The program expires August 31, 2007, so there is no funding in the budget baseline. *The Administration recommends renewing the MILC program at the current 34% payment rate for FY2008, and then gradually reducing the rate to 20% over the coming six years.*

A July 2004 USDA study on the "Economic Effects of U.S. Dairy Policy and Alternative Approaches to Milk Pricing" concludes that "current dairy programs are limited in their ability to change the long-term economic viability of dairy farms" and that the MILC program contributes to dairy surpluses and reduces the farm milk price. Some observers would argue that the price support program (which removes milk from the market) and the

Title I: Commodity Programs 9

MILC program (which encourages more production) appear to be working at cross-purposes. Why does USDA recommend a continuation of current dairy policy?

Since the MILC program's inception, large dairy farms have contended that the 2.4 million lb. payment limit (a herd size of about 125 cows) is biased against them, given that 2.4 million lbs. represents a small portion of their production. What is the federal government's response to their concerns?

Under our current WTO trade obligations, the aggregate measure of support for dairy is based on how much higher the domestic support price is set above a fixed world reference price, and this imputed subsidy is applied to all domestic milk production. Using this formula, the WTO views the aggregate measure of support for the dairy price support program to be more than $4.5 billion annually (even though federal outlays are well below $1 billion), and classifies it as "amber box" (the most trade-distorting category). The current U.S. proposal in the Doha Round is to reduce its total amber box support from the current $19.1 billion to $7.6 billion. With dairy support contributing so much toward the proposed new maximum, did the USDA consider proposing an alternative to current policy that is decoupled from price and production?

Since the MILC program's inception in 2002, it has provided total counter-cyclical payments of $2.4 billion over five marketing years (2002-2006). USDA estimates that the 10-year total cost of extending and revising the MILC program under its proposal is $793 million (FY2008-FY2017). Why is the cost estimate so far below historical expenditures on the program? Is it due to projections for improved dairy market conditions, or do proposed revisions to the program significantly reduce expenditures? Why are changes to the other commodity support programs being proposed that would save money, but the status quo would be maintained for dairy policy that will cost additional money over baseline?

The current MILC program calculates payments based on current monthly production levels. The Administration's farm bill proposal would base payments on 85% of the three-year average of milk marketed during FY2004-FY2006, instead of on current production. The proposal states that such a change would make the MILC program consistent with other farm bill counter-cyclical programs. What is the rationale for making this change to the MILC program? What are the trade implications of making this change (i.e., will it be enough to consider the program decoupled and the payment categorized as "green box")?

Sugar

Support for raw cane sugar and refined beet sugar are mandatory under the 2002 farm bill at $0.18 and $0.229 per pound respectively. The prices are guaranteed by nonrecourse loans available to cane processors and beet refiners. However, because the United States is a net importer of sugar, market prices usually can be maintained above the mandatory support levels by limiting supplies through import barriers.

The United States is authorized a global tariff rate quota of 1.256 million short tons under WTO rules that is allocated among sugar exporting countries around the world. Mexico separately is limited to shipping 276,000 short tons through calendar year 2007, and then has open access to the U.S. market under the North American Free Trade Agreement (NAFTA). An additional supply control feature of the law allows for imposition of domestic marketing allotments on U.S. sugar, but only when imports are less than 1.532 million short tons. The suspension of allotments when imports increase was adopted in the 2002 farm bill at the urging of the domestic sugar industry.

With no limit on Mexican sugar imports after 2007, and several bilateral free trade agreements adopted or in process that would allow in more sugar, the United States is faced with the likelihood of imports exceeding 1.532 million short tons. The imports in excess of 1.532 million short tons likely will go under price support loan and eventual forfeiture to the CCC at an estimated cost of $1.107 billion over the next 10 years, according to USDA.

The farm bill proposal would continue the sugar support program and the current nonrecourse loan rates, but would eliminate the provision in current law requiring the Secretary to suspend marketing allotments when sugar imports are projected to exceed 1.532 million short tons. This change in the law is projected to save $1.107 billion over 10 years.

If current law is extended, how much sugar would the CCC likely acquire under the loan program and what would be the disposal outlets?

If the government savings of $1.107 billion from the imposition of sugar allotments were translated into reduced revenues for sugar farmers, what would that amount to?

Has USDA examined a direct payments program on sugar base acres as an alternative to current policy? What would a direct payment program likely cost if it were designed to leave sugar producers in an equivalent net cash income situation, assuming they have flexibility to earn revenue from other crops?

SPECIAL COTTON COMPETITIVENESS PROVISIONS

The 2002 farm bill included three provisions to enhance cotton export competitiveness but protect domestic textile mills from high prices. Step 1 allowed for a downward adjustment under specific circumstances in the adjusted world price (AWP, which is analogous to the posted county price for grains) for upland cotton, which increases the loan deficiency payment (LDP) to producers. Step 2 mandated offsetting payments to exporters and domestic users of cotton when U.S. prices were higher than world prices so that the buyers were not disadvantaged by buying U.S. cotton. The Step 2 provision for upland cotton was repealed by Congress following a ruling that it violated the WTO Agreement on Agriculture. Step 3 allowed for a special additional import quota for upland cotton when high world prices for U.S. cotton and Step 2 export subsidies created tight supplies for domestic mills.

The USDA's farm bill proposes to eliminate Step 1 because it has been used infrequently. When it has been used the result has been increased costs for cotton LDPs. Step 3 would be eliminated because its purpose has disappeared with the elimination of Step 2. Additionally, Step 2 for extra long staple (ELS) cotton would be eliminated because it is analogous to Step 2 for upland cotton, which was eliminated after being found in violation of WTO rules.

How much has been spent (in total and per pound) on Step 2 for ELS cotton under the current farm bill?

Would elimination of Step 1 and Step 3 have any adverse consequences for upland cotton producers or domestic textile mills?

PLANTING FLEXIBILITY LIMITATIONS

Current law (first adopted in the 1996 farm bill and continued by the 2002 farm bill) prohibits, except in certain limited circumstances, the planting of fruits, vegetables, and wild rice on program crop base acres. Violation of this restriction results in the loss of direct and counter-cyclical payments. With the exception of these commodities, farmers do have planting flexibility on base acres. Practically, this means that corn base acres can be planted to any other subsidized crop and vice versa, but not to fruits or vegetables. The limitation was put in place because producers of unsubsidized, but high value, specialty crops objected to likely competition from subsidized land.

For purposes of meeting its WTO obligations, the United States has considered direct payments on base acres to be minimally production- and trade-distorting because they are decoupled from production decisions and market prices. Consequently, direct payments have been reported to the WTO as "green box" and not counted against the $19.1 billion limit on "amber box" trade-distorting subsidies. A WTO ruling on the U.S. cotton program reasoned that the planting flexibility restriction does not meet criteria for decoupled income support. *The USDA farm bill proposal would eliminate the restriction on planting fruits, vegetables, and wild rice on base acres in order to make direct payments fully compliant with the WTO green box rules.*

What impact would elimination of the base acre planting restriction have on fruit, vegetable, or wild rice producers? If there are impacts, where would they be most severe?

Does this situation demonstrate that U.S. agriculture is an integrated sector that cannot be divided up easily into independent components for special treatment, or not?

Does this interaction between farms and commodities lend support to a policy of whole farm revenue insurance instead of a patchwork of subsidies and rules that generate inefficiencies and inequities?

CROP BASE RETIREMENT

Under current law, cropland that is converted to nonagricultural uses does not retain eligibility for commodity program subsidies. However, it is possible to convert cropland to nonagricultural uses without losing base acre benefits that were tied to the converted cropland. USDA points out that an owner of two farms can transfer the base acreage benefits from a farm being sold to another farm being retained so long as the farm receiving the transferred base is sufficiently large in size to accommodate the increased base. Another example is the retention of all of the base even though part of the farm is sold. *The USDA's farm bill proposal would proportionally reduce base acreage whenever all or part of a farm is sold for nonagricultural uses.*

Was the base acreage retirement provision made to achieve equity among farmers or as a disincentive to convert cropland to nonagricultural uses? Could it be argued that the proposal encourage farmers to sell entire farms to developers instead of small parts of the farm and thereby accelerate the conversion of cropland to nonagricultural uses?

CONSERVATION ENHANCED PAYMENT OPTION

The currently operating Conservation Security Program (CSP) provides technical and financial assistance to participants who address, at a minimum, water and soil resources concerns, through conservation, protection, and improvement. Larger payments are made to participants who address additional resource concerns on their entire operation. All farmers are eligible to apply for the program. However,
limited funding of $502 million (between FY2004 and FY2006) has constrained the program to 14% of the nation's 2,119 watersheds, and many farmers have found the administrative burden to be excessive. To date, 19,291 contracts have enrolled 15,411,134 acres into CSP in 298 watersheds.

The Secretary proposes that farms with program crop base acres be offered a "conservation enhanced payment" equal to 10% of the commodity program direct payment for adopting conservation and environmental practices equivalent to the Progressive Tier requirement of the Conservation Security Program (CSP). Farmers electing this option would forgo their counter-cyclical payments and marketing loan benefits for the duration of the 2007 farm bill.

Is this Conservation Enhanced Payment Option effectively a pilot effort to convert "amber box" commodity programs into "green box" payments?

Some farms with commodity base acres now have CSP contracts, plus they receive direct and counter-cyclical payments as well as marketing loan benefits. Would the Conservation Enhanced Payment Option create two categories of CSP participants with substantially different benefits?

Some observers may ask why would a farmer give up the commodity program safety net of potentially large counter-cyclical payments and marketing loan benefits for the certainty of only a 10% increase in the direct payment. What is the federal government's response?

Are any farmer costs associated with achieving the Progressive Tier requirement, and, if so, would those costs be more than covered by the 10%

"enhancement" or would there be other federal assistance for those expenses?

CONTINUING WTO COMPLIANCE

The 2002 farm bill includes "circuit breaker" authority for the Secretary of Agriculture to make adjustments in domestic commodity support expenditures when needed to comply with Uruguay Round Agreements. The U.S. annual limit on trade distorting (amber box) subsidies is $19.1 billion. *The Secretary has proposed that the circuit breaker authority be modified to accommodate any new agreements from the Doha round of negotiations or other agreements concluded under the auspices of the World Trade Organization (WTO).*

Has the circuit breaker authority ever been invoked by the Secretary in order to avoid exceeding the $19.1 billion U.S. amber box limit?

Has the United States ever exceeded its amber box limit? What has been the size of U.S. amber box subsidies each year since the Uruguay Round Agreements were adopted?

Chapter 2

TITLE II: CONSERVATION

Before the 1985 farm bill, few conservation programs existed and only two, the Agricultural Conservation Program and Watershed and Flood Prevention Operations, would be considered large by today's standards. In total, conservation programs were funded at less than $1 billion annually. The current conservation portfolio includes more than 20 distinct programs with annual spending of about $5.2 billion. Most are enacted through recent farm bills with mandatory funding supplied by USDA's Commodity Credit Corporation (CCC). The 2002 farm bill authorized large increases in mandatory funding for several agricultural conservation programs. The two largest, the Conservation Reserve Program (CRP) and the Environmental Quality Incentives Program (EQIP), make up almost 55% of the $5.2 billion in current annual spending.

The Administration's proposed 2007 farm bill has outlined an overall increase in funding for agricultural conservation programs, which the Administration estimates is $7.825 billion over the 10-year current services baseline of $48.698 billion. Much of this additional funding is attributed to an increase in the proposed consolidated Environmental Quality Incentives Program (EQIP) and an increase in the acreage limit for the Wetlands Reserve Program (WRP). Many of the Administration's proposed changes would consolidate existing programs, with the goal of increasing administrative efficiencies and reducing participant confusion.

Several program consolidation changes are proposed. What level of savings can be expected by these consolidations and are there any specific plans for using those "savings?"

Given the continued growth of the conservation effort, what additional evaluation measures, if any, are planned to keep Congress informed about accomplishments and spending efficiencies?

Based on the Administration's budget plan and the farm bill proposal, there may appear to some to be inconsistencies within working lands conservation programs. The Administration proposed increasing funding for CSP and EQIP by a combined $475 million annually in the farm bill proposal. The FY2008 budget proposal, meanwhile, proposes to reduce both CSP and EQIP. EQIP is authorized at $1.27 billion and the President's budget requests $1.0 billion (a $270 million reduction). CSP is estimated by the CBO at $451 million and the President's budget requests $316 million (a $135 million difference). Would the proposed increases in mandatory conservation programs authorized by the farm bill supersede the cuts in those same programs if the Administration's budget were adopted? How should the differences between the farm bill and the budget proposal be interpreted?

ENVIRONMENTAL QUALITY INCENTIVES PROGRAM (EQIP)

EQIP offers agricultural producers cost-share payments, technical assistance, and incentive payments to implement conservation practices on private working-lands. Three sub-programs are implemented through EQIP: the Ground and Surface Water Conservation Program (GSWC), the Klamath Basin Program, and Conservation Innovation Grants (CIG). The GSWC program targets areas with extensive agricultural water needs to achieve a net savings in water consumption. While the Klamath Basin program is similar in nature, it is limited to a single basin that straddles the California-Oregon border. CIG, a grant program, is intended to foster the development and adaptation of innovative conservation approaches.

Other conservation programs utilize cost-sharing mechanisms similar to EQIP, including the Wildlife Habitat Incentives Program (WHIP), which focuses on developing and restoring wildlife habitat on all land; the Agricultural Management Assistance (AMA) Program, which seeks to mitigate risks through diversification and resource conservation practices; and the Forest Land Enhancement Program, which addresses resource concerns on private forest lands.

Citing duplication in eligibility requirements, regulations, policies, applications, and administrative actions, the Secretary recommends consolidating existing cost-share programs (EQIP, GSWC, WHIP, AMA, Forest Land Enhancement Program, and the Klamath Basin Program) into a newly designed EQIP program. The proposal also would create a Regional Water Enhancement Program (RWEP) to address water quality and quantity issues on a regional scale. The CIG program would receive additional funding. This newly constructed program would receive an increase of $4.25 billion over the OMB 10-year current services baseline.

Interest in participating in many conservation programs has been high, leading to a large backlog of unfunded applications. EQIP alone reported 32,633 unfunded applications, worth more than $636 million, in FY2006. With such a large application backlog in EQIP along with the other proposed programs for consolidation, would the additional authorized funding be used primarily in addressing the application backlog? Would USDA have the workforce capacity to handle the workload added by this increase in funding?

USDA's farm bill spending estimates show that EQIP grows modestly until it reaches its full spending authority in 2014. This would occur after a new five-year farm bill has expired. Is there an explanation as to why EQIP, a program with a standing backlog of applications would require seven years to "ramp up" to its fully authorized spending level?

Some of the programs proposed for consolidation have very specific programmatic purposes and eligibility requirements targeted at specific resource concerns on different components of the landscape (WHIP focuses on wildlife habitat on all lands, while GSWC focuses on water quantity on only agricultural lands). How would specific resource problems be targeted in the absence of specialized programs? If the consolidation of these programs requires a single definition of eligible lands, what definition would the Department prefer?

CONSERVATION SECURITY PROGRAM (CSP)

The Conservation Security Program (CSP) provides technical and financial assistance to participants who address, at a minimum, water and soil resources concerns, through conservation, protection, and improvement. It was widely touted as the first conservation entitlement and the first "green payments" program when enacted in 2002. The program operates with three conservation and funding tiers, Tier III being the highest. Larger payments are made to participants who address additional resource concerns on their entire operation (Tier III). There are currently four components to CSP financial assistance payments: (1) stewardship, or base payments for the number of acres enrolled, (2) maintenance payments for existing conservation practices, (3) cost-share payments for new practices, and (4) enhancement payments for conservation effort and additional activities beyond a prescribed level. All farmers in eligible watersheds may apply for the program. Limited funding of $502 million (between FY2004 and FY2006) has constrained the program to 298 of the nation's 2,119 watersheds, and many farmers state that they have found the administrative burden to be excessive. To date, more than 19,000 contracts (averaging 800 acres) have enrolled 15.4 million acres into CSP.

The Secretary proposes the following adjustments: elimination of stewardship, maintenance, and cost-share financial assistance payments; consolidation of three tiers into two tiers; creation of a ranking system in place of the current watershed approach; and expansion of funding to $8.5 billion during FY2008-FY2017.

How would the proposed changes alter the pattern and scale of participation?

Will the proposed Conservation Enhancement Option in the commodity provisions of Title I overlap or replace CSP for producers of program crops?

Current law limits technical assistance spending in conjunction with CSP to 15% of the program cost. Some observers claim this level is inadequate. No recommendation was made to repeal the 15% limitation on technical assistance in current law. How can CSP be successfully implemented with what some would argue is such a small level of technical assistance?

PRIVATE LANDS PROTECTION PROGRAM

The proposed Private Lands Protection Program would consolidate three existing programs. The Farmland Protection Program (FPP) purchases conservation easements to limit nonagricultural uses of land. The Grasslands Reserve Program (GRP) seeks to restore and protect rangeland, pastureland, and other grassland while continuing grazing sustainability. The Healthy Forest Reserve Program (HFRP) addresses forest land that provides habitat for threatened and endangered species.

Citing common goals and unique eligibility requirements, regulations, policies, applications, and administrative actions, the Secretary recommends consolidating these three existing easement programs (FRPP, GRP, and HFRP) into a new private lands protection program. This proposal also would increase funding by an additional $900 million over the 10-year baseline for this new program.

Some consider FPP a working lands program that keeps farming viable, while GRP is more closely related to a land restoration program. How would these fundamental differences be resolved in a consolidated program?

CONSERVATION RESERVE PROGRAM (CRP)[1]

The Conservation Reserve Program (CRP) and Conservation Reserve Enhancement Program (CREP) remove active cropland into conservation uses, typically for 10 years, and provide annual rental payments (based on the agricultural rental value of the land) and cost-share assistance. Conversion of the land must yield adequate levels of environmental improvement to qualify (environmental benefits index). CRP is the largest land retirement program, with spending of $1.828 billion in FY2005. The total program acreage is limited to 39.2 million.

The Secretary recommends reauthorization of CRP with an enhanced focus on lands that provide the most benefit for environmentally sensitive lands. Priority would be given to whole-field enrollment for lands utilized for energy-related biomass production. Biomass would be harvested after nesting season and rental payments would be limited to income forgone or costs incurred by the participant to meet conservation requirements in those years biomass was harvested for energy production.

With cellulose conversion technology in its infancy, what is the rationale behind subsidizing cellulose production at this time?

The proposal may appear to some observers to have two conflicting components with regard to CRP. If it is desirable to focus CRP on multi-year idling of more environmentally sensitive lands, some may inquire why the harvesting of biomass on those lands is being imposed. Could this harvesting conflict with the conservation purpose of the program?

If it is decided that high demand for commodities dictates that less land should be in the CRP, how would priorities be set for land to be released?

WETLANDS RESERVE PROGRAM (WRP)

The Wetlands Reserve Program (WRP) and Wetlands Reserve Enhancement Program (WREP) provide technical and financial assistance to private landowners to restore and protect wetlands. WRP has a current enrollment of 1.89 million acres with an annual authorized new enrollment cap of 250,000 acres. The 2002 farm bill authorized a total enrollment cap of 2.275 million acres.

The Secretary recommends the consolidation of WRP with the floodplain easement program of the Emergency Watershed Program and an increase in the enrollment cap to 3.5 million acres. This increase in acreage would equal an estimated $2.125 billion increase over the current services 10-year baseline of $455 million.

By increasing the acreage cap for WRP and proposing a continuation of the CRP acreage cap, what consideration is given to the 25% county cropland enrollment cap on CRP and WRP?

Similar to CRP, what effect would increased enrollment of wetlands have on local economies? Would it be possible to achieve the objectives of wetlands conservation under a working-lands program rather than a cropland retirement program?

How would these changes affect the President's no-net-loss goal?

SOD SAVER

The 1985 farm bill included the first conservation compliance requirement for farmers to participate in certain USDA programs. The conservation compliance provision for highly erodible land, also known as Sodbuster, created disincentives to farmers who produced annually tilled agricultural commodities on highly erodible cropland without adequate erosion protection. Conservation compliance for wetlands, also known as Swampbuster, created disincentives to farmers who produced annually tilled agricultural commodities or made possible the production of agricultural commodities on land classified as wetlands.

The Administration recommends broadening the conservation compliance provisions to include new "Sod Saver" rules that would create a disincentive to converting grassland (rangeland, and native grasslands, not previously in crop production) into crop production. *Sod Saver would make* all newly converted grasslands permanently ineligible for commodity support and other USDA programs (including other conservation programs). The suggested date for this provision to go into effect is not stated in the Administration's proposal. The Administration scores this proposal at zero budgetary impact.

What is the basis for the Sod Saver recommendation? Is Sodbuster not working as well as anticipated? Is the concern of some conservationists that Sodbuster has not been aggressively enforced a valid concern or not?

Was an alternative considered in the Sod Saver provision to allow for approved conservation systems that could provide for a reduction in soil erosion, similar to the conservation compliance for highly erodible land (Sodbuster) and wetland conservation compliance (Swampbuster)?

Sod Saver may not prevent the conversion of grasslands to cropland when crop prices are high. Once cultivated, there could be some off-site damages. Yet Sod Saver will preclude any federal assistance to address these problems. What is the rationale behind permanently prohibiting conservation assistance on this converted land?

Presumably there will be monitoring and enforcement costs associated with Sod Saver. What agency would have the administrative responsibility and what would be the estimated costs?

CONSERVATION ACCESS FOR BEGINNING AND LIMITED-RESOURCE FARMERS

First recognized as requiring special attention in the 2002 farm bill, beginning and limited-resource farmers are provided with additional incentives in conservation programs through various funding mechanisms and targeted initiatives. The largest incentive directed toward beginning and limited-resource farmers is the increase in cost-share payments (up to 90% of the cost to implement the practice can be paid by NRCS) in EQIP. Other programs such as the Conservation Innovation Grants and Farm Protection Program also have initiatives directed toward beginning and limited-resource farmers.

The Administration is recommending that 10% of farm bill conservation financial assistance be reserved for beginning farmers and ranchers, as well as socially disadvantaged producers. Flexibility is also recommended to allow the Secretary to reallocate the reserve funds if the money goes unused.

The Administration states that this proposal would have no effect on the current services baseline.

What portion of current conservation program participants meet the description of beginning and limited-resource farmers?

How much is currently spent on beginning and limited-resource farmers, and how does that relate to the 10% of financial assistance being proposed?

Would implementation of this provision mean that some commercial farmers would then be unable to participate in conservation programs? If so, to what extent would this occur?

MARKET-BASED FUNDING

The Secretary recommends that $50 million, over a 10-year period, be available to develop uniform standards for environmental services, establish credit registries, and offer credit audit certification services to encourage new private sector environmental markets to supplement existing conservation programs.

Are details available on how the market-based approach would work and what the return on this public investment would be? What are some models or examples?

How was $50 million determined to be an appropriate 10-year budget?

EMERGENCY LANDSCAPE RESTORATION PROGRAM

Both the Emergency Watershed Protection Program (EWP) and the Emergency Conservation Program (ECP) provide disaster assistance to private landowners through discretionary technical and financial assistance appropriated by Congress. The EWP, administered by the Natural Resource Conservation Service (NRCS), focuses on impairments to watersheds caused by natural disasters. It works through local sponsors such as neighborhood associations, cities, counties, and conservation districts. The ECP, administered by the Farm Service Agency (FSA), focuses on emergency water conservation measures in periods of severe drought on farmland, and also provides assistance to rehabilitate farmland damaged by all natural disasters.

Citing confusion and frustration by citizens responding to natural disasters, *the Secretary is recommending that the EWP and ECP be consolidated into a new Emergency Landscape Restoration Program.* Funding for this new program would be discretionary, as is the current funding stream for EWP and ECP.

Since the Emergency Watershed Protection (EWP) program and the Emergency Conservation Program (ECP) are administered by two different agencies, which agency would administer the new consolidated Emergency Landscape Restoration Program?

Currently, land eligibility is very different between the two programs. Should the consolidation of these programs include a single definition of eligible lands; if so, what should be the definition? If different definitions of eligible land are maintained, does this affect the goals of consolidation?

Chapter 3

TITLE III: TRADE

Farm bills typically authorize multi-year funding for USDA agricultural trade programs (direct export subsidies, export credit guarantees, foreign food aid, and export market development) and address new issues that have arisen as U.S. agricultural exporters seek to sell their products overseas.

TECHNICAL ASSISTANCE FOR SPECIALTY CROPS

USDA's Foreign Agricultural Service (FAS) is responsible for promoting U.S. agricultural exports, including advocating on behalf of U.S. agricultural interests in foreign capitals and in international organizations as disputes arise. Funding for FAS staff and expenses to accomplish this and related objectives is provided through the annual appropriations process. In addition, the 2002 farm bill authorized an initiative — Technical Assistance for Specialty Crops Program (TASC) — to fund projects that address sanitary and phytosanitary (SPS) and technical barriers related to specialty crops. TASC is a mandatory program, meaning that it is funded by tapping the Commodity Credit Corporation's (CCC's) borrowing authority. The 2002 farm bill authorized a funding level of $2 million each year for FY2002-FY2007 ($12 million total).

The Administration's farm bill proposals seek to expand funding for TASC. Funding would be phased in at $4 million in FY2008, $6 million in FY2009, $8 million in FY2010, and $10 million in each subsequent year through FY2015 (for a multi-year total of $68 million). USDA also proposes to increase the maximum allowable annual project award from $250,000 to $500,000 and allow more flexibility to allow project timeline extensions.

USDA argues that additional flexibility would allow for the acceptance of larger, multi-disciplinary projects that result in better quality proposals from eligible participants and improved assistance to specialty crop growers.

Why does USDA request that this new initiative be funded on a mandatory basis, using CCC's borrowing authority? Is the appropriations mechanism a more suitable approach for funding this?

What evidence is there that TASC projects have resulted in the elimination of SPS and/or technical barriers to trade in specialty crops and that they have contributed to increased U.S. agricultural exports?

MARKET ACCESS PROGRAM

The Market Access Program (MAP) assists in the creation, expansion, and maintenance of foreign markets for primarily U.S. agriculture high-value products. This program funds the U.S. government's share of the cost of overseas marketing and promotional activities with non-profit U.S. agricultural trade associations, U.S. agricultural cooperatives, nonprofit state-regional trade groups, and small U.S. businesses. Activities include consumer promotions, market research, trade shows, and trade servicing. About 60% of MAP funds typically support generic promotion activities (i.e., non-brand name commodities or products), and about 40% support brand-name promotion (i.e., a specific company product).

The 2002 farm bill authorized MAP's six-year funding level at $875 million (FY2002-FY2007), rising from $100 million in FY2002 to $200 million in each of FY2006 and FY2007. MAP is a mandatory program, funded by the CCC.

The Administration recommends increasing annual MAP funding by $25 million each year ($250 million over 10 years). The additional funding would be used to "address the inequity between farm bill program crops and non-program commodities," and represents one of several recommendations offered in USDA's farm bill proposal to assist specialty crop producers.

Why is the Administration considering an increase in MAP funding, when past (FY2006 and FY2007) budget proposals called for cutting the authorized level in half, from $200 million to $100 million?

How much of MAP's funding already assists specialty product producers and firms, and how much of a difference would a $25 million annual increase make?

How does USDA gauge the impact of MAP? What evidence is there that discrete MAP promotion activities in particular country markets have resulted in an increase in U.S. agricultural exports?

SANITARY AND PHYTOSANITARY (SPS) ISSUES GRANT PROGRAM

Reportedly, developing and developed countries are increasingly using unscientific SPS standards as non-tariff barriers to U.S. agricultural products. These take the form of plant and animal health restrictions to protect their domestic agricultural sectors against outside competition. Examples often cited include biotechnology restrictions, maximum residue standards, and restrictions on U.S. beef (such as those imposed by South Korea and Japan) due to BSE (mad cow disease).

The Administration's farm bill proposes to establish a new grant program to address SPS issues for all commodities (mandatorily funded by the CCC at $2 million each year, or $20 million over 10 years). This program would allow for new or expanded focus on such issues as foreign governments' acceptance of antimicrobial treatments; wood packaging material; irradiation; biotechnology; science-based maximum residue level standards; and testing procedures and controls for mycotoxins. Grants would fund projects that address SPS barriers that threaten U.S. agricultural exports, by reducing the need to hire technical staff on a permanent basis, involving the private sector in assisting USDA in solving technical problems, commissioning scientific reports on targeted issues, and making more use of outside technical experts to address these types of barriers.

What are the most significant SPS barriers that currently affect U.S. agricultural exports? What is the status of U.S. efforts to address these specific barriers?

How much of Foreign Agricultural Service (FAS) resources already are tapped to address SPS issues?

How would an SPS Grant Program reinforce U.S. efforts to eliminate SPS barriers in international standard setting forums or in WTO dispute settlement?

Why does USDA request that this new initiative be funded on a mandatory basis, using CCC's borrowing authority? Is the appropriations mechanism a more suitable approach for funding this?

INTERNATIONAL TRADE STANDARD SETTING ACTIVITIES

Reportedly, countries have increasingly resorted to technical trade barriers that have no scientific basis in order to restrict imports of U.S. agricultural products. One U.S. effort to counter this trend is to become more involved in international bodies that establish and harmonize multilateral food, plant, and animal safety standards —frequently referred to as sanitary and phytosanitary (SPS) rules. Such organizations include the Food and Agriculture Organization (FAO), the *Codex Alimentarius*, the International Plant Protection Convention, and the World Animal Health Organization (known by its French acronym of OIE).

Acknowledging that the U.S. government lacks sufficient resources to ensure that its views on SPS issues are fully heard, USDA is requesting authority and mandatory funding of $15 million over 10 years ($1.5 million annually) to enhance USDA staff support at international standard setting organizations.

Funding would be used to close the compensation gap for senior level U.S. staff placed in these organizations to influence decision making (e.g., ensure that standards are properly designed and implemented to avoid unwarranted trade barriers), and to cover the cost of up to four professional officers who would specifically focus on supporting U.S. SPS priorities.

How many USDA staff already serve to represent U.S. interests and perspective at the FAO and these three international bodies? How long has this function been carried out?

The Administration's proposal notes that the lack of U.S. funding for staff support has led the FAO to take a "more Eurocentric approach" in its work, "which may be in conflict with U.S. objectives." Can the Administration elaborate on what this approach means, and comment on how that perspective affects U.S. efforts to reduce or eliminate technical trade barriers?

TECHNICAL ASSISTANCE FOR TRADE DISPUTES

The number of U.S. agricultural trade disputes has increased in recent years. This has prompted commodity groups and agribusiness firms to seek recourse under U.S. trade remedy laws to address potential unfair competition in the domestic U.S. market, and to work with USDA and the U.S. Trade Representative (USTR) to try to resolve cases considered under the World Trade Organization's trade dispute process. The process of pursuing a dispute case is usually complex, lengthy, and costly, particularly for smaller groups and agricultural industries with limited resources.

The Administration's farm bill requests broad discretionary authority to provide enhanced monitoring, technical assistance, and analytical support to agriculture groups with limited resources, if the Secretary of Agriculture determines this would benefit U.S. agriculture. This would enable USDA to direct available resources to assist smaller agricultural groups and industries affected by unfair foreign trade practices and to pursue trade dispute cases on their behalf.

Why is this authority needed, in light of USDA's statement that it already helps out in trade dispute cases by providing legal and analytical support, often working with USTR?

In what instances would a program of technical assistance for trade disputes have enabled small groups or agricultural industries to pursue a case in WTO dispute settlement or have affected the outcome of cases that have been pursued?

What criteria would USDA use in exercising such broad authority to determine which groups should be deemed eligible for this type of assistance?

TRADE CAPACITY BUILDING AND AGRICULTURAL EXTENSION PROGRAMS IN STRATEGICALLY IMPORTANT COUNTRIES

In recent years, USDA has worked with the Departments of State and Defense, and the National Security Council, in Afghanistan and Iraq to provide technical assistance in support of efforts to revitalize the agriculture sectors of both countries. Such assistance was provided through existing agricultural extension programs, but USDA did not receive direct funding for such activities.

The Administration proposes providing $20 million in mandatory funding over 10 years ($2 million annually), through CCC's borrowing authority, to expand agricultural extension and food safety programs in fragile countries. This would be part of the U.S. government's efforts to meet future development assistance needs in unstable areas, such as Sudan or Somalia. USDA's role would be to engage in agricultural reconstruction and extension efforts, targeted towards those who are dependent upon agriculture for food and employment.

Would funding agricultural extension efforts in fragile countries facilitate U.S. agricultural trade with them? How would trade capacity building function in very unstable circumstances?

What are USDA's efforts in trade capacity building elsewhere in the world? What level of existing resources does USDA already tap for such activities?

EXPORT CREDIT PROGRAMS AND FACILITY GUARANTEE PROGRAMS

USDA administers four export credit guarantee programs to facilitate sales of U.S. agricultural exports. Under these programs, private U.S. financial institutions extend financing to foreign buyers of agricultural products, with the Commodity Credit Corporation (CCC) guaranteeing repayment in case of borrower default. The CCC guarantee facilitates a more favorable interest rate and a longer repayment period. Eligible countries are those that USDA determines can service the debt. Use of guarantees for foreign aid, foreign policy, or debt rescheduling purposes is prohibited.

The Short-Term Export Credit Guarantee Program (GSM-102) guarantees short-term financing (up to three years). Separately, the Intermediate-Term Export Credit Guarantee Program (GSM-103) guarantees intermediate-term financing (up to 10 years). The Supplier Credit Guarantee Program (SCGP) guarantees deferred payment sales (usually up to 180 days). The Facilities Guarantee Program (FGP) is to improve or establish the handling, marketing, storage, or distribution facilities for U.S. commodities in emerging markets.

The 2002 farm bill authorized up to $6.5 billion annually for the FY2002-FY2007 period for these guarantee programs. Of this amount, $1.0 billion is targeted to "emerging markets" — countries in the process of becoming commercial markets for U.S. agricultural products. The statute gives USDA's Foreign Agricultural Service (FAS) the flexibility to determine the allocation between short and intermediate term programs. The actual level of guarantees approved each year depends on market conditions

Title III: Trade

and on the demand for financing by eligible countries. Program activity has declined over the last three years because of less demand for guarantees and administrative steps taken in July 2005 to bring the programs into conformity with a World Trade Organization (WTO) ruling that found them to be prohibited export subsidies. In FY2006, USDA approved almost $1.4 billion in guarantees, down from $2.6 billion in FY2005 and $3.7 billion in FY2004. The budget outlay impact of guarantees ($142 million in FY2006) is small because it reflects only administrative costs and the subsidy associated with the loans approved each year.

The Administration proposes statutory changes to reform these guarantee programs in light of the WTO ruling and to ensure they remain WTO-compliant. These include (1) removing the current 1% cap on fees collected under the GSM-102 program, (2) eliminating the specific authority for the GSM-103 program, (3) terminating the SCGP (because of the large number of loan defaults (totaling $227 million) and evidence of fraud), and (4) revising the FGP to attract additional users who commit to purchase U.S. agricultural products. The first three proposed changes are not expected to have any budgetary impact, according to USDA. The cost of changes to the FGP would be minor (almost $2 million each year).

To what degree would higher loan guarantee fees diminish user participation in the short-term GSM-102 program?
What impact on program activity is anticipated with the proposed lifting of the current 1% fee?
What is the prospect for USDA collecting on the more than $200 million in SCGP loan defaults?
What explains declining interest among countries in using USDA loan guarantees to finance their agricultural imports?
What fundamental changes are occurring in worldwide commodity financing that may warrant revisiting the role that credit guarantees can play in facilitating U.S. agricultural exports?

EEP AND TRADE STRATEGY REPORT

USDA established the Export Enhancement Program (EEP) in 1985 to help U.S. commodities compete with other countries, primarily the European Union, that subsidized their exports. Used extensively through the late 1990s to challenge unfair trade practices and maintain market share in targeted countries, EEP has been inactive in recent years. The 2002 farm bill established an annual program level of $478 million, the maximum allowed under the Uruguay Round export subsidy reduction commitments.

The 2002 farm bill requires the Secretary of Agriculture to consult with Congress every two years and to prepare a Global Market Strategy report that identifies growth opportunities overseas for U.S. agricultural exports. The administrative costs of preparing one report are about $250,000. USDA further notes that this requirement duplicates its existing United Export Strategy and Country Strategy programs, which use real-time market analysis and global intelligence, and are more timely.

USDA proposes to repeal EEP and the Global Market Strategy report mandate, pointing to program inactivity and the report's redundancy. It argues that because both no longer serve valuable purposes, the proposed changes would allow USDA to focus staff and financial resources to priority issues. USDA notes that EEP's repeal would also be consistent with the U.S. objective to eliminate the use of export subsidies worldwide.

How does the Administration estimate no budgetary impact from this proposal while it justifies the change in terms of financial savings?

If there is not a successful conclusion this year to the Doha Round of WTO trade negotiations, is the United States placed at a disadvantage (i.e., would it lose a tool to assist agricultural exports in the future) if it unilaterally repeals the law authorizing an export subsidy program?

What are examples of how USDA's existing United Export Strategy and Country Strategy programs are more useful to U.S. exporters and policymakers?

CASH AUTHORITY FOR EMERGENCY FOOD AID

USDA provides food aid abroad through the P.L. 480 program, also known as Food for Peace; the Food for Progress Program; the McGovern-Dole International Food for Education and Child Nutrition Program; and Section 416(b) of the Agricultural Act of 1949. The 2002 farm bill authorized all of these programs through FY2007, except for Section 416(b), which is permanently authorized by the Agricultural Act of 1949. The 2002 farm bill also reauthorized a commodity reserve of wheat and other commodities typically used as food aid (renamed the Bill Emerson Humanitarian Trust), which can be used, under certain circumstances, to provide P.L. 480 food aid; and created the McGovern-Dole program as a new food aid program.

Funding for the P.L. 480 programs (Title I direct credits, and Title II grants) is provided through the annual appropriations process. Title I provides for long-term, low interest loans to developing and transition countries and private entities to purchase U.S. agricultural commodities. The use of Title I credits has declined over time, and totaled $123 million in FY2006. Title II provides for the donation of U.S. agricultural commodities to meet emergency and non-emergency food needs. The law mandates an annual minimum tonnage level of 2.5 million metric tons. In recent years, appropriators have set the funding level between $1.6 and $1.7 billion. The Food for Progress Program, funded directly by the CCC ($131 million in FY2006), provides commodities to support countries that have made commitments to expand free enterprise in their agricultural economies. The McGovern-Dole program uses commodities and financial and technical assistance to carry out preschool and school food for education programs and maternal, infant and child nutrition programs in foreign countries. The 2002 farm bill mandated CCC funding of $100 million in FY2003 and authorizes appropriations of "such sums as necessary" from FY2004 to FY2007. The FY2006 program level was $97 million. Donations through the Section 416(b) program are entirely dependent on the availability of commodities acquired by the CCC in its price support operations. The Emerson Trust provides emergency food relief when U.S. supplies are short or to meet unanticipated need.

Under current law, Title II of P.L. 480 may only be used to purchase and ship U.S. agricultural commodities to meet food needs overseas. The Administration points out that this stipulation has precluded the use of Title II to procure food quickly enough, or resulted in the United States not being able to provide food or provide it late. It argues that authority is needed to quickly meet emergency needs in the most effective way possible, such as using cash to provide immediate relief until U.S. commodities arrive or to fill in gaps in the food aid pipeline. It notes that U.S.-sourced food aid typically takes four months or longer to arrive where needed, compared to days or weeks when commodities can be purchased locally. The same case has been made by the Administration in the FY2006 and FY2007 budget proposals, but Congress has rejected the idea both times. *The Administration's farm bill proposes to authorize the use of up to 25% of P.L. 480 Title II funds for the local or regional purchase and distribution of emergency food to respond more quickly to assist people threatened by a food security crisis. There is no direct budgetary impact associated with this proposal.*

Are there examples from the past where the proposed food aid authority could have been used to more quickly provide assistance and thereby helped alleviate tragic food emergencies?

Has USDA identified developing countries where food could be purchased to help with any currently existing emergencies?

Are current U.S. food aid program resources sufficient to meet outstanding needs in trouble spots around the world?

Has the Administration considered proposing additional funding for the McGovern-Dole program — providing food aid to youngsters in schools, particularly girls, in order to also meet broader objectives of fighting poverty?

How will the Administration's focus on emergency food aid affect the availability of food aid for development purposes?

How will using U.S. funds to purchase overseas commodities rather than U.S. commodities affect the willingness of U.S. groups (private voluntary organizations, farm organizations, commodity groups, maritime industry) to support the U.S. food aid program? Could the Administration's proposed approach result in a lower availability of U.S. food aid to meet humanitarian needs?

In: The USDA 2007 Farm Bill Proposal
Editors: J. Womach et al.

ISBN: 978-1-60456-813-4
© 2008 Nova Science Publishers, Inc.

Chapter 4

TITLE IV: NUTRITION

FOOD STAMP PROGRAM: WORKING POOR AND ELDERLY

Retirement Savings

Under standard federal rules, "defined benefit" retirement plans, "401(k)" plans, and several other types of retirement savings arrangements are now excluded as assets when determining food stamp eligibility. But other retirement/savings plans like Individual Retirement Accounts (IRAs) and Simplified Employer Pension (SEP) plans are not disregarded. Also under current law, states exercising an option to conform food stamp rules to those of their Temporary Assistance for Needy Families (TANF) program can expand the standard federal disregard to match their TANF rules, and TANF and Supplemental Security Income (SSI) recipients are automatically eligible for food stamps (essentially making food stamp asset eligibility rules irrelevant for them). Effectively, current rules governing the treatment of retirement savings/plans vary noticeably by state, type of applicant, and type of plan.

Retirement/education savings that are counted are included, with other countable assets, under the Food Stamp program's general limit on assets — $2,000, or $3,000 if the household includes an elderly or disabled member. Other countable assets generally include liquid resources like cash or assets readily converted to cash (but not household belongings/furnishings), some illiquid resources (e.g., real property not producing income, but not a household's home), and, to varying degrees (by state and type of vehicle), the value of household-owned vehicles.

The Administration proposes to disregard all retirement savings and plans as assets when judging food stamp eligibility. Its stated purposes are to reinforce federal policy encouraging retirement savings and to end the penalty that counting them imposes on those experiencing a temporary need for food assistance. Some critics argue that the initiatives should go further toward liberalizing the treatment of assets (e.g., raising or abolishing dollar limits on assets, standardizing the disregard for vehicles). Others contend that the current system of state TANF-based options and automatic food stamp eligibility for TANF/SSI recipients provides enough flexibility to address any need to liberalize the food stamp asset eligibility test.

The Administration estimates costs for its retirement savings proposal at $548 million over five years and $1.305 billion over 10 years. The proposal also is included in the Administration's FY2008 budget package.

How many food stamp applicants does the Administration estimate its proposed disregard for retirement savings will affect?

The Administration's proposal for retirement savings appears to deal with formal, tax-recognized situations. What about money put aside by poor households for retirement that is not part of a formal plan?

In general, eligible food stamp households must have countable assets not exceeding $2,000 (or $3,000 for the elderly or disabled). Although these dollar limits apply to fewer types of assets than when they were established, they have not been changed in over 20 years. Has the Administration considered raising (or indexing) the dollar limits on countable assets?

Food stamp eligibility rules governing counting vehicles as assets are complex and vary significantly by state and vehicle type (e.g., whether it is work-related), similar to rules for retirement savings. Has the Administration considered standardizing and simplifying these rules, as it proposes for retirement savings?

With fewer types of assets being counted in judging eligibility for food stamps (and other programs) and more flexibility being given to states, is keeping a food stamp asset test administratively cost-effective? Is it true that a number of states already effectively eliminated asset tests for TANF benefits?

Reimbursement for Work-Related Expenses (Pilot Project)

Current food stamp rules provide federal 50% matching for state support for the work-related expenses of food stamp recipients in work/training programs; states choose who is covered, what expenses will be reimbursed, and generally what the reimbursement will be. Employed recipients receive no similar support — although they may increase their benefits by claiming a "deduction" for work-related dependent care costs (see the separate proposal on the treatment of dependent care expenses) and a "deduction" for 20% of any earnings to cover taxes and work expenses and can, in some cases, get separate aid through state child care and TANF programs and income tax provisions.

The Administration proposes to establish a three-year, three-state, $3 million pilot project under which states would be allowed to pay (with 50% federal matching) for work-related expenses (other than dependent care costs) of households with earnings from employment. The stated purpose is to test an idea that might further strengthen the Food Stamp program's role in supporting work and moving individuals and families to self-sufficiency. USDA would define what expenses would be covered (child care costs would not be allowed) and could place a limit on the time recipients could be aided in the project. Critics question whether the dollar and covered-expense restrictions placed on the project effectively limit the usefulness of any results.

How can a pilot program that pays a limited range of work expenses for employed food stamp recipients, and is restricted to $3 million, produce meaningful results? Does the $3 million include evaluation costs? Will there be control groups? Is there experience from comparable TANF initiatives to indicate whether this type of support is potentially productive?

Is it the Administration's intention to propose that this type of work-expense support for the employed be made a regular feature of the Food Stamp program if the pilot proves a success?

Dependent-Care Expenses

Food stamp law takes dependent care costs related to work or education into account when determining eligibility and benefits. It does this by allowing households to "deduct" these costs from countable income — up to certain limits. As a result, households with these costs are more likely to be

eligible, and more important, are given a larger food stamp benefit; benefits generally
increase by 30 cents for each dollar of disregarded income. Dependent care cost deductions are "capped" at $200 a month for each child under age two and $175 a month for all other dependents, thereby limiting the extent to which these costs affect food stamp eligibility and benefits.

The Administration proposes to eliminate the current caps on expense deductions for dependent care costs used when calculating food stamp eligibility and benefits. The stated purpose is to help working families. Critics argue for the need to go further in recognizing the effect high non-food living expenses — like shelter costs — have in eroding the value of food stamp benefits.

The Administration estimates costs for this proposal at $20 million over five years and $42 million over 10 years.

How many households does the Administration expect to be affected by its proposal to lift the dollar caps on dependent-care expense deductions?

A dollar cap, albeit an indexed one, also exists for shelter-expense deductions. Has the Administration considered lifting it to increase benefits for those with very high shelter expenses?

College Savings Plans

Current food stamp policy allows a disregard of college (postsecondary education) savings plans as assets to the extent that they are determined to be "inaccessible." It also permits states to exclude college savings plans when conforming their food stamp rules to their TANF or Medicaid policies. As a result, state agencies must make individual determinations as to the accessibility of education savings in order to judge whether to disregard them — unless they have a TANF or Medicaid rule that disregards them and have chosen to apply that rule to food stamps.

Education savings that are counted are included, with other countable assets, under the Food Stamp program's general limit on assets — $2,000, or $3,000 if the household includes an elderly or disabled member. Other countable assets generally include liquid resources like cash or assets readily converted to cash (but not household belongings/furnishings), some illiquid resources (e.g., real property not producing income, but not a household's home), and, to varying degrees (by state and type of vehicle), the value of household-owned vehicles.

The Administration proposes to disregard Internal Revenue Service (IRS)-approved postsecondary/college education savings plans as assets when judging food stamp eligibility. The stated purposes are to reinforce federal policy encouraging savings for education, to end the penalty that counting them imposes on those experiencing a temporary need for food assistance, and to simplify program administration. Some critics argue that the initiatives should go further toward liberalizing the treatment of assets (e.g., raising or abolishing dollar limits on assets, standardizing the treatment of vehicles). Others maintain that the current rule allows for a disregard where the savings have been set aside for education and are truly inaccessible for living expenses.

The Administration estimates costs for its education savings proposal at $8 million over five years and $18 million over 10 years. The proposal also is included in the Administration's FY2008 budget proposal.

How many food stamp applicants does the Administration estimate its proposed disregard for education savings will affect?

The Administration's proposal for education savings appears to deal with formal, tax-recognized situations. What about money put aside by poor households for education that is not part of a formal plan?

In general, eligible food stamp households must have countable assets not exceeding $2,000 (or $3,000 for the elderly or disabled). Although these dollar limits apply to fewer types of assets than when they were established, they have not been changed in over 20 years. Has the Administration considered raising (or indexing) the dollar limits on countable assets?

Food stamp eligibility rules governing counting vehicles as assets are complex and difficult to administer, similar to rules for education savings. Has the Administration considered standardizing and simplifying rules for disregarding vehicles as assets, as it proposes for education savings?

With fewer types of assets being counted in judging eligibility for food stamps (and other programs) and more flexibility being given to states, is keeping a food stamp asset test administratively cost-effective? Is it true that a number of states already effectively eliminated asset tests for TANF benefits?

Combat-Related Military Pay

Combat-related military pay has been disregarded as income in the Food Stamp program through provisions of appropriations laws since the FY2005 agriculture appropriations act. This proposal would make the disregard part of permanent food stamp law.

For a number of years, the Defense Department has offered a Family Assistance Supplemental Allowance to military families who might qualify for food stamps. Its purpose is to increase their income and make participation in the Food Stamp program unnecessary; however, very few families have chosen to take this option.

The Administration proposes to disregard, in permanent law, combat-related military pay as income when determining food stamp eligibility and benefits. The stated purpose is to permanently remove a potential penalty on military families. Critics might contend that use of the Defense Department's special allowance program would be a better way to deal with this issue.

The Administration estimates costs for its military pay proposal at $5 million over five years and $10 million over 10 years. This proposal also is included in the Administration's FY2008 budget package.

How many military families does the Administration estimate would be affected by the proposed disregard of combat-related pay for food stamp purposes?

Should these families be participating in the Defense Department's Family Assistance Supplemental Allowance program instead of the Food Stamp program? How big is the supplemental allowance program?

Has the Administration considered similar treatment for civilian employees deployed in combat areas?

FOOD STAMP PROGRAM: STREAMLINING AND MODERNIZING PROPOSALS

Rename the Food Stamp Program

The Food Stamp program got its name when it was originally established in 1939. At the time, actual "stamps" were used. Blue and orange stamps were issued to recipients — one color representing the recipient's dollar contribution and the other the federal government's subsidy (which was usable only for surplus commodities). When used, the stamps were actually pasted into booklets by the participating grocer, and the booklets (when full) were then redeemed by the retailer for cash. The original program was closed down in 1943, after World War II had effectively eliminated surplus food production.

When the program was revived in the 1960s, the old name also was revived —even though stamps were not used. Instead, participants received paper "coupons" of various denominations that were used like cash to purchase food. This lasted until the recent switch to the use of debit-card-like electronic benefit transfer (EBT) cards to deliver benefits.

The Administration recommends changing the program's name to the Food and Nutrition program in recognition of the changes in how food stamp benefits are delivered and the program's role in improving nutrition.

USDA opened up renaming of the program for public comment a few years ago and received many ideas. Could a compendium of those ideas be provided?

What is the estimated cost of switching to a new name for the program?

How many states now use a name other than Food Stamp program to identify their EBT-based program?

"Stamps" have not been used since the early 1940s, yet the program has kept the name. Is it an option to keep the existing name and let states call the program what they wish?

"De-Obligate" Food Stamp Coupons as Legal Tender

Food stamp benefits are now delivered using debit-card-like electronic benefit transfer (EBT) cards, not paper food stamp coupons. The EBT system has been in place nationwide for two years. However, some coupons issued before the transition to EBT systems have still not been redeemed, and the Administration would like to "get them off the books," saving redemption costs (both the value of the coupons themselves and the cost of handling them). *The Administration proposes to "de-obligate" food stamp coupons still in circulation, making them no longer usable (redeemable).* It estimates net savings from this proposal at $2 million over five years and $7 million over 10 years.

What is the dollar value of unredeemed coupons still outstanding?
What time deadline for redeeming outstanding coupons does the Administration envision?

Prohibit Certain State Claims Against Recipients; Collect Over-issuances from States

Current law requires states to pursue collection of over-issued food stamp benefits from recipients and former recipients — however they are caused. Any collections generally are turned over to the federal government (which pays the cost of program benefits), but states may keep a portion of collections in cases where the over-issuance was not caused by the state agency's actions. To an extent, states themselves also are liable to the federal government for over-issued benefits and losses caused by cases of state agency negligence or fraud.

The Administration proposes to (1) prohibit states from pursuing claims against recipients for over-issued food stamp benefits in the case of "widespread systemic errors" (e.g., computer system failures/flaws) and (2) require states to pay the USDA for over-issuances in such cases. The stated purposes for advancing its proposals are to promote program integrity and fair treatment of recipients and to encourage caution and careful planning when implementing new computer-related administrative systems. However, a number of states are pursuing initiatives that encompass the expanded use of computers and online interaction between applicants/recipients and state agencies, and some critics are concerned over how extensively any new authority to require state payments might be used and its potential "chilling effect" on state efforts to improve administration.

How will the Administration define "widespread systemic errors" and calculate over-issued benefits for the purpose of holding recipients harmless and mandating that states pay the cost of the over-issuances they might cause? What about under-issuances and improper denials caused by these "widespread systemic errors?"

What type of new authority to collect from states is the Administration asking for? Does USDA not have enough authority already to collect from states in cases of over-issued benefits? Is this authority not being used in the Colorado case that the Administration cites in the rationale for its proposal?

Will the threat of a new requirement that states pay over-issuance costs in cases of computer system flaws and other systemic problems dampen the current trend toward state innovation in administering food stamps by increasing the use of computers and online interactions between applicants/recipients and state agencies, or not?

FOOD STAMP PROGRAM: IMPROVE PROGRAM INTEGRITY

Limit Categorical Eligibility

Under current food stamp law, states may grant categorical (automatic) food stamp eligibility to households receiving Temporary Assistance for Needy Families (TANF) cash aid, services, or both —effectively accepting TANF eligibility decisions as to financial eligibility for food stamps. In some cases, particularly where only *services* are provided, the household may have financial resources (income/assets) significantly above those normally allowed for food stamps.

The Administration is concerned that states can, in effect, "game" the categorical eligibility option and make households eligible for food stamps by simply providing minimal services financed with TANF funds. *Its proposal would restrict categorical (automatic) eligibility for food stamps to those who receive cash benefits under state Temporary Assistance for Needy Families (TANF) programs* (on the premise that they are subject to stricter eligibility tests than those getting TANF-funded services).[2] On the other hand, critics point out that, among those categorically eligible, it is most often working households with relatively high non-food expenses (for shelter, dependent care) who actually qualify for a significant food stamp benefit, and that many of the households that would be penalized are those who have worked their way off cash welfare and are only receiving child care services to help them keep their job. They also note that there would be added administrative costs and a significant side effect — some households losing their categorical food stamp eligibility would, as a result, lose their food-stamp-participation-based categorical eligibility for free school meals for their children.

The Administration estimates savings from this proposal at $611 million over five years and $1.360 billion over 10 years. It also is included in the Administration's FY2008 budget package and was advanced as part of the FY2007 budget presentation.

Is it the Administration's intent to exclude any household receiving only TANF-funded services from categorical eligibility for food stamps? Would the proposal exclude working households getting child care aid? Those getting job training? What TANF-funded services do those excluded by the Administration's recommendation receive?

How many households would be affected by the categorical eligibility limitation proposal? Is this proposal the primary source of budget savings in the Administration's package of farm bill proposals for nutrition assistance programs? Without this cost-saving change, would the Administration continue to support the recommendations it has made that have significant projected costs?

In how many states is overly expansive categorical eligibility a problem? What types of services are provided in these states to confer categorical food stamp eligibility?

Could a state "get around" the Administration's proposal by providing a minimal cash payment instead of a service?

Has the Administration estimated the added administrative costs that states (with a 50% federal match) would bear for conducting regular food stamp eligibility determinations for those losing categorical eligibility who choose to apply through regular program rules, particularly checking on assets held by applicants?

Imposing Fines on Retailers

Under current policy, the use of fines as penalties on retailers violating food stamp rules (e.g., selling non-food items for food stamp benefits) is restricted to certain instances where the retailer can show extensive efforts to educate employees (and the owner was unaware of the violation) or where disqualification would cause hardship to food stamp recipients. In most cases, it is USDA policy to impose disqualification (for varying periods), whether the violation is minor or major. *The Administration argues that it needs more flexibility to respond to the seriousness of a retailer's violation and would like more authority to impose fines in lieu of disqualification for minor violations and new authority to impose fines in addition to disqualification for major violations.* Critics are concerned that the extent of the problem (beyond anecdotal cases) is not clear, that USDA has not aggressively used its existing authority to substitute fines in minor cases, and that authority to impose fines in addition to disqualification in major cases might clash with separate provisions of law that impose court-ordered monetary penalties on retailers convicted of felonies/misdemeanors.

The Administration estimates savings from these proposals at $5 million over five years and $10 million over 10 years.

How many, and what types of, cases point to the need for new authority to impose fines instead of disqualification in cases of relatively minor retailer violations of food stamp rules?

Why, specifically, are current authorities for the use of fines on retailers in cases of minor violations not sufficient?

Seizure of Retailers' Food Stamp Receipts

Under current law, retailers accused of trafficking violations can continue to operate (and potentially continue any fraudulent activities) while enforcement actions are taking place — "even if those violations are particularly egregious." *The Administration proposes to allow the USDA, in "certain egregious trafficking cases," to seize retailers' food stamp receipts prior to settlement in cases where expedited action is warranted.* The stated purpose is to increase the effectiveness of USDA enforcement actions. As stated by the Administration, "trafficking retailers [would be] hurt more quickly where it matters — in their pocketbooks. This proposal increases effectiveness by immediately stopping the flow of funds that allow retailers to continue to finance their fraudulent activities." Critics are worried about how this new authority would be framed and used and whether it goes too far and possibly "pre-judges" accused retailers.

How would the proposed authority to seize retailers' food stamp receipts be framed? What protections for accused retailers is envisioned?

What type of "egregious" cases would the new seizure authority be designed and used for? In what types of cases (and how many) would it have been used in recent years?

Recipient Disqualification for Selling Food

Food stamp law requires disqualification for those who traffic in food stamp benefits (i.e., those who exchange the value of benefits on their food stamp EBT card for cash or ineligible items). However, selling/trading the actual food purchased with food stamp benefits for cash or other consideration is not cause for disqualification. *The Administration proposes to disqualify those who exchange food purchased with food stamp benefits for cash.*

How will the Administration enforce a rule disqualifying those who exchange food purchased with food stamp benefits for cash? Is this proposal only intended to deal with egregious cases? What about cases where food obtained with food stamp benefits is exchanged for something other than cash?

Does the Administration envision its proposal also making the exchange of food purchased with food stamp benefits a felony or misdemeanor subject to fines or imprisonment — like trafficking under current law — or only cause for disqualification?

Penalties on States for High Negative Action Error Rates

The Food Stamp program has a "quality control" (QC) system under which state administration of the program is measured for the extent of erroneous determinations — that is, annual "error rates" are computed for overpayments to eligible and ineligible recipients and underpayments to eligible recipients. For FY2005 (the most recent year for which QC figures are available), the national overpayment rate was 4.53% of food stamp benefit dollars, and underpayments were valued at 1.31% of food stamp benefits paid out. States with consistently (over three consecutive years) high error rates may be assessed fiscal sanctions. These sanctions are calculated as a portion of the cost of improper payments and the value of proper payments not made — above certain allowable thresholds. At present, eight states are at risk of a sanction or have a sanction liability; current actual liabilities total $3.6 million. States also may receive payment accuracy bonus payments for overpayment and underpayment error rates that are very low or greatly improved; for FY2005, 10 states received bonus payments totaling $24 million.[3]

In addition to overpayments and underpayments, the food stamp QC system measures the extent to which states improperly deny, suspend, or terminate food stamp applicants/recipients (i.e., annual negative action error rates are calculated for each state). In FY2005 (the most recent year for which these figures are available), the national average negative action error rate was 6.91%, and nine states had rates 50% or more above the national average (six of them were in their second consecutive year). States with high negative action error rates are not subject to fiscal sanctions. However, they may receive bonus payments for very low or greatly improved negative action rates; for FY2005, six states received bonuses totaling $6 million.

Overpayment and underpayment error rates have been dropping in recent years; in FY2005, they were at a historic low. On the other hand, negative action error rates have been rising. The Administration proposes to assess states a financial penalty if the state has a negative action error rate above the national average for two consecutive years. It appears that the penalty would be a dollar amount equal to 5% of the federal share (normally 50%) of a state's food stamp administrative costs. The Administration's stated purposes in advancing its proposal to impose penalties for high negative action error rates are to promote program integrity and correct eligibility determinations for applicants. Critics are concerned that it reopens the extensively negotiated 2002 farm bill agreement with states and advocates that reformed the QC system and imposes an overly large penalty for high negative action error rates, without adequate grounds for doing so. The Administration estimates savings from this proposal at $57 million over five years and $166 million over 10 years.

When compared to fiscal sanctions assessed for food stamp overpayments and underpayments, the proposed sanction for high rates of improper negative actions, like mistaken eligibility denials, is very large. Administration estimates indicate they will average over $15 million a year. Is there a reason for this substantial difference?

In the 2002 farm bill, the previous practice of assessing sanctions as a reduction in the federal share of state administrative costs was abandoned in favor of sanctions as a proportion of the dollar value of improperly paid or unpaid benefits. Why has the Administration chosen to use this sanction method for high rates of negative actions? Is it possible that a cut based on administrative spending could exacerbate the problem? Does the Administration intend to include cuts in federal matching payments for state costs like nutrition education and work and training programs in the proposed sanction?

State Financial Liability for High Error Rates

States at risk of a fiscal sanction for consistently high error rates as measured by the food stamp QC system (see the discussion above of the Administration's proposal for penalties for high negative action error rates) may meet a portion of the sanction by investing (using unmatched state funds) in federally approved improvements to the administration of the Food Stamp program. Citing a need to boost program integrity and strengthen the

QC system, *the Administration proposes to eliminate the option permitting states to invest in administrative improvements as an alternative to paying part of their QC fiscal sanction.* As with the proposal for penalties for high negative action error rates, critics contend that this recommendation unnecessarily re-opens the 2002 agreement that revamped the QC system without a sufficient rationale. The Administration estimates minimal cost savings from this proposal.

Has the USDA had any problems with states' use of the current option to invest in administrative improvements in lieu of paying fiscal sanctions? Have states not fulfilled their administrative improvement promises? Has administration been significantly enhanced by these efforts?

How many states have taken advantage of the option to pay for administrative improvements instead of paying the USDA? How much money was involved and what types of enhancements were made?

FOOD STAMPS: IMPROVING HEALTH THROUGH NUTRITION EDUCATION

Recognize Nutrition Education as a Component of the Program

Current policy authorizes federal 50% matching payments for state nutrition education efforts for food stamp recipients and the potentially eligible low-income population — as an allowable administrative cost. USDA also funds the cost of providing nutrition education materials and technical assistance related to nutrition education. *The Administration proposes to add specific language to the Food Stamp Act referring to nutrition education as an approved activity under the Food Stamp program.*

Does the Administration's proposal envision funding any nutrition education activities not now supported?

Pilot Obesity Initiative

The Administration is concerned over substantial indications that obesity among Americans is rising. At present, the USDA nutrition programs support nutrition education activities and have a few features directed at combating obesity. For example, the Food Stamp program pays half the cost of state nutrition education efforts among food stamp recipients and other low-income households; the USDA provides nutrition education materials and makes grants for nutrition education initiatives directed at schoolchildren; school meal program meal patterns are being revised and schools are required to design "wellness policies" that address obesity concerns; and the WIC program includes a major, mandatory nutrition education component.

In addition to these efforts, the Administration's proposal calls for competitive grants to develop and test ways of addressing obesity in the low-income population — with evaluations of the results. This would be accomplished with a five-year, $20 million per year "USDA Initiative to Address Obesity among Low-Income Americans." According to the USDA, ideas that might be tested include point-of-sale incentives for the purchase of fruit and vegetables by food stamp recipients, grants to connect food stamp shoppers with farmers' markets, and integrated communication and education programs to promote healthy diets and physical activity.

How would the pilot obesity initiative be coordinated with existing child nutrition and WIC program efforts and projects supported by the Department of Health and Human Services?

What are the reasons given for USDA to embark on a separate new grant initiative?

THE EMERGENCY FOOD ASSISTANCE PROGRAM (TEFAP)

Permanent State TEFAP Plans

Under current law, states must submit plans of operation and administration for USDA approval every four years; plans may be amended at any time with USDA approval. State plans designate the state agency responsible for distributing TEFAP commodities, set out the state's plans for distributing commodities, and set forth eligibility standards for participating agencies and individual recipients. *The Administration proposes to make all TEFAP state plans effectively permanent and require that states only submit revisions that are warranted by changes in state TEFAP operations or rules.*

This is very close to the pattern for state plans in other USDA nutrition programs, and the Administration argues that the current once-every-four-years rule is burdensome on state TEFAP agencies. Critics, on the other hand, question whether a requirement to resubmit a state plan once every four years is really that burdensome on states (as opposed to USDA officials) and note that TEFAP state plans are more important than those in other USDA nutrition programs because states have almost total control over program rules and operations. They also point out that a complete plan review and re-submission every four years (if done conscientiously by the state and the USDA) can result in important program improvements and that other programs' state plan requirements were changed to revisions-as-warranted from previous once-a-year requirements (not once every four years).

Have states called for a change in the rules governing submission of state TEFAP plans? Is the current requirement for new state plans every four years more of a burden on the USDA or state agencies?

If the problem is a burden on USDA plan reviewers, would staggering submission of state plans be an appropriate solution?

Selecting TEFAP Local Organizations

Under current policies, states have complete control over the selection of local organizations that receive and distribute the TEFAP foods allocated to each state (including those groups that act as conduits to end providers like food pantries and soup kitchens, and end providers themselves). The Administration is concerned that this situation can lead to many of the same organizations participating year after year with little concern over how efficiently or effectively they are delivering services, unless significant administrative problems occur. In order to encourage the entry of new distributing organizations that might operate more efficiently and could charge lower fees to end providers, *the Administration proposes to require that states re-compete contracts with TEFAP distributing organizations at least once every three years.* It contends that failure to have a periodic competitive solicitation process results in a barrier to "certain local organizations, including faith-based organizations, that wish to participate in TEFAP." Critics are concerned over how a requirement for competitive selection would work given the widely varying nature of state TEFAP programs. They also question why *all* contracts need to be renewed (recompeted) so often, whether there are potentially enough serious competitors to make the three-year competition process worthwhile, and the cost of running competitive solicitations and changing distributing organizations. And they point to the potential for confusion among recipients when distribution systems change.

Does the Administration's proposal for re-competing contracts for TEFAP distributing agencies mean that all contracts will potentially be subject to termination and reassignment at the same time every three years? Will there be some staggering of contract renewals?
What federal rules for competitive solicitations are envisioned?
Have local organizations that are not now part of the TEFAP distribution network asked for competitive solicitations?

FOOD DISTRIBUTION PROGRAM ON INDIAN RESERVATIONS (FDPIR)

Increased Administrative Funding

The Administration, in consultation with participating Indian tribal organizations, is in the process of substantially revising the method for allocating federal payments for administrative costs for FDPIR. The new method would be more closely tied to participants served. The dollar amount to be spent on administrative costs and the allocation of federal payments for them are *not* specified in the underlying law governing the FDPIR.

To speed and support implementation of a revised allocation (i.e., ease the negative effects for tribal organizations that would lose money under a new allocation), the Administration is asking for increased FDPIR administrative funding of $26-$27 million over 10 years. Critics question whether this initiative belongs in the farm bill and how it would be crafted given that it deals with matters not now covered in FDPIR law.

Has a decision on a new method for allocating FDPIR administrative payments been made? If not, does the Administration know the amount of new funding needed (and the years in which it will be needed)?

The new funding the Administration is asking for is described as ensuring that there would be sufficient money so that any change in the current allocation method would allow all tribal organizations to "continue to receive their current allotments or a modest increase depending on their level of participation." What would it cost to ensure that, under the new allocations, all tribal organizations are held harmless (including inflation increases)? Will the Administration's proposal include the details of the new allocation method? Should this be placed into law?

Could the Administration's goal be achieved through the regular appropriations process?

Food Stamp/FDPIR Disqualification Policies

The FDPIR is a program distributing federally donated foods that is operated in lieu of food stamps on Indian reservations where the tribal organization opts for it. Individuals cannot participate in both food stamps and the FDPIR at the same time. Under current policy, those disqualified (e.g., for fraud) from the Food Stamp program are automatically disqualified from participation in the FDPIR (following the food stamp disqualification rules). On the other hand, those disqualified from the FDPIR are not similarly disqualified from food stamps.

The Administration proposes to change food stamp law to specifically disqualify from food stamps those disqualified from the FDPIR in order to promote program integrity and consistent eligibility/disqualification rules. Critics question whether new provisions of law are needed to accomplish this.

Why can't food stamp disqualification of those disqualified from the FDPIR be accomplished by a change in food stamp regulations using the disqualification authorities provided in Section 6(b) and Section 6(h) of the Food Stamp Act?

SENIOR FARMERS' MARKET NUTRITION PROGRAM (SFMNP)

Disregarding SFMNP Benefits in Other Assistance Programs

The SFMNP provides once-a-year vouchers (typically worth $20-$30) to low-income seniors; these vouchers are used at participating farmers' markets and roadside stands to buy fresh produce. *The Administration proposes to require that the value of SFMNP vouchers be disregarded in federal and state means-tested public assistance programs.* This change is intended to make treatment of SFMNP vouchers consistent with the treatment of other nutrition assistance benefits (e.g., the WIC Farmers' Market Nutrition program, child nutrition program benefits, food stamps).

Do any public assistance programs now count the value of SFMNP vouchers as income or financial resources — specifically, other nutrition assistance programs like food stamps?

Prohibiting Sales Taxes on SFMNP Purchases

The Administration proposes to prohibit states from participating in the SFMNP if state or local sales taxes are charged on food purchased with SFMNP vouchers. This recommendation is intended to make treatment of SFMNP vouchers consistent with the treatment accorded other nutrition assistance benefits (e.g., the WIC Farmers' Market Nutrition program, food stamps).

Do any states or localities now charge sales taxes on SFMNP voucher purchases? Does the Administration expect any state to pull out of (or not apply for) the program if sales taxes on vouchers are barred?

PROMOTING HEALTHFUL DIETS IN SCHOOLS

School Food Purchase Survey

The most recent data on school food purchases are a decade old. The Administration proposes to require a $6 million survey of foods purchased by schools for their meal services, once every five years.

It maintains that, in addition to getting information on fruit and vegetable purchases, its proposed periodic surveys would help USDA efforts to (1) provide guidance to schools in implementation of upcoming new rules intended to conform school meal patterns to the most recent Dietary Guidelines for Americans, (2) better manage the commodities procured by the USDA for distribution to schools, and (3) assess the economic effect of school food purchases on various commodity sectors. Critics ask whether this belongs in the farm bill.

Could the Administration's goal be achieved through the regular appropriations process?

New Funds for Fruit and Vegetable Purchases for Schools

In recent years, USDA has acquired an average of over $300 million a year in fruit and vegetables for schools. About $50 million is purchased and distributed through the "Department of Defense Fresh Program," which supplies fresh fruit and vegetables to schools under contract with the USDA. In response to calls for an increase in the quality of USDA-provided commodities, *the Administration proposes to provide an additional $50 million a year for the purchase of fruit and vegetables specifically for the School Lunch program — above acquisitions under any other authority.*

Some of this new spending could be through added dollars for the Defense Department Fresh Program. Critics are concerned that the Administration may not be asking for a large enough increase in fruit and vegetable purchases and that its farm bill proposals are silent on potential expansion of a small ($15 million) existing fresh fruit and vegetable program operating in some 400 schools located in 14 states and on 3 Indian reservations.

What is the Administration's position on expansion of the fresh fruit and vegetable program initiated in the 2002 farm bill and later expanded?

How would the $50 million a year in new fruit and vegetable purchases requested by the Administration be distributed among schools?

Will the proposed $50 million for fruit and vegetable purchases be mandatory funding? Could the Administration's goal be reached through the regular appropriations process?

Section 32 Fruit and Vegetable Purchases[4]

"Section 32" is a permanent appropriation that since 1935 has earmarked the equivalent of 30% of annual customs receipts to support the farm sector through a variety of activities. Today, most of this appropriation (now approximately $7 billion yearly) is transferred to the U.S. Department

Title IV: Nutrition

of Agriculture (USDA) account that funds child nutrition programs. However, a smaller — but still significant — amount of Section 32 money is set aside each year to purchase non-price-supported commodities directly and provide them to schools and other feeding sites. Some of these purchases are "entitlement" commodities that are required to be made under school lunch law. Others are "bonus" commodities, acquired by USDA through emergency surplus removal activities. The total value of both types of commodities now exceeds $900 million per year. The purchases are made by USDA's Agricultural Marketing Service (AMS). Included within these combined ("mandated" and "bonus") Section 32 totals, fruit and vegetable purchases over the last five years have averaged $308 million per year, according to USDA.

In order to promote healthy diets, **USDA** proposes to increase purchases of fruits and vegetables using Section 32 authority by at least $200 million per year, and $2.75 billion over 10 years. However, critics are concerned over actual extent of any new fruit and vegetable purchases and their effect on Section 32 support for other commodities.

Documents detailing the budget effect of the Administration's farm bill proposals indicate no score (no new spending) for its Section 32 recommendation. How does the Administration propose to cover the cost of these increased fruit and vegetable purchases?

If new spending would not be created, which activities or food purchases would be reduced to pay for these increases? For example, Section 32 is now also used to purchase animal products including meats, poultry, and seafood. Would the Administration's proposal result in fewer purchases of these products? If not, why?

USDA routinely has funds remaining in the Section 32 account at the end of each fiscal year, which are "carried over" into the next fiscal year to be used in Section 32. What is this level of unobligated funds, on average, and is there any intent to reduce the size of this carryover to pay for new fruit and vegetable purchases? If so, could that leave less carryover in future years?

Does this proposal call for any new legislative authority, and if not, how can Congress be assured that the initiative would be carried out by future Administrations?

How does USDA currently determine what proportions of its Section 32 commodity acquisitions go to various domestic nutrition programs, and how would it do so for the proposed increases?

How does this proposal differ from the separate Administration initiative providing for $50 million yearly in other new fruit and vegetable purchases for domestic nutrition programs? How would it be funded?

Does the Department need broad legislative authority to administer Section 32 programs, particularly "bonus" surplus removals?

In: The USDA 2007 Farm Bill Proposal
Editors: J. Womach et al.
ISBN: 978-1-60456-813-4
© 2008 Nova Science Publishers, Inc.

Chapter 5

TITLE V: CREDIT

The Administration proposes three revisions to the permanently authorized farm loan programs of the USDA's Farm Service Agency (FSA). FSA is a lender of last resort, providing direct and guaranteed loans to farmers unable to secure credit elsewhere. The general intention of the farm bill proposal is to enhance loan availability for beginning and socially disadvantaged classes of farmers and ranchers, and to increase the maximum size of individual direct loans, which effectively have been reduced by inflation. The cost of these changes against the budget baseline is zero because the programs are funded by annual discretionary appropriations. The statutory changes in eligibility and loan size may affect the distribution of program benefits and how far a dollar of appropriation goes, but appropriators will continue to control the actual level of spending.

First, the Administration proposes to target more of the FSA direct loan portfolio to beginning and socially disadvantaged farmers. Currently, the law requires a certain percentage of the loan authority to be reserved for beginning farmers and ranchers for a specific length of the fiscal year, and funds are disbursed across states by expected need. After the targeting period ends, any remaining funds are pooled across states and allocated to other qualified farmers. *The Administration proposes to double the targeting percentage for direct operating loans from 35% to 70%, and increase the targeting of direct farm ownership loans from 70% to 100%. New re-pooling procedures at the end of the targeting period would redistribute funds first to targeted groups of farmers in other states before other farmers.*

Second, the Administration proposes to enhance the beginning farmer down payment program to make it easier for beginning and socially disadvantaged farmers to buy property. It would (a) lower the interest rate charged from 4% to 2%, (b) eliminate the $250,000 cap on the value of property that may be acquired, (c) decrease the producer contribution from 10% to 5%, (d) defer payments for the first year, and (e) add socially disadvantaged farmers to the list of eligible applicants.

Third, the Administration proposes to raise the current $200,000 borrower limit on direct farm ownership loans and $200,000 limit on direct farm operating loans to a combined $500,000 limit on both types of loans. The current limits were established in 1984 and 1978, respectively, and have been eroded in terms of purchasing power by inflation in the price of land and inputs. Limits on guaranteed loans were increased in 1998, indexed for inflation, and combined across ownership and operating loans.

The proposed $500,000 combined limit on direct farm operating and farm ownership loans is not indexed for inflation. However, the limit on guaranteed loans dating from 1998 is indexed for inflation. What is the rationale for indexing guaranteed loans and not direct loans?

Farmers may have more flexibility with the combined $500,000 cap, but the total is nonetheless only slightly higher than the current $400,000 total across the two types of loans. Given the increase in land prices and input costs since the mid 1980's, is a 25% increase in the combined loan cap sufficient?

The 2002 farm bill required a study of the effectiveness of the delivery of USDA's direct and guaranteed loan program. The issue was whether the direct loan program was still needed, given shifts in many different government loan programs toward guaranteed loans, including at FSA. The Administration's FY2008 budget for rural development calls for cutting direct loans in the rural housing program. Why does the USDA believe direct farm loans are still necessary but not direct rural housing loans?

What has been USDA's experience with the pilot program to guarantee contract land sales as established under the 2002 farm bill (7 U.S.C. 1936)? The program was authorized as a pilot through FY2007, and was to guarantee loans made by a private seller of a farm to a qualified beginning farmer on a contract land sale basis. How would USDA rate the success of this program? Why is USDA not requesting its reauthorization?

Does the Administration have a position on expanding the lending authority of the Farm Credit System (FCS), a policy FCS supports but commercial bankers oppose?

In: The USDA 2007 Farm Bill Proposal ISBN: 978-1-60456-813-4
Editors: J. Womach et al. © 2008 Nova Science Publishers, Inc.

Chapter 6

TITLE VI: RURAL DEVELOPMENT

Three agencies established by the Agricultural Reorganization Act of 1994 (P.L. 103-354) are responsible for USDA's Rural Development mission area: the Rural Housing Service (RHS), the Rural Business-Cooperative Service (RBS), and the Rural Utilities Service (RUS). An Office of Community Development provides community development support through Rural Development's field offices. The mission area also administers the rural portion of the Empowerment Zones and Enterprise Communities Initiative, the Rural Economic Partnership Zones, and the National Rural Development Partnership.

RURAL CRITICAL ACCESS HOSPITALS

The Critical Access Hospital Program was created by the 1997 Balanced Budget Act (P. L. 105-33) as a safety net device, to assure Medicare beneficiaries access to health care services in rural areas, and to create incentives to develop local integrated health delivery systems, including acute, primary, emergency and long-term care. Assistance for medical care facilities and other essential community facilities has been provided under USDA Rural Development's Community Facilities program.

The Administration proposes mandatory funding of $1.6 billion in guaranteed loans and $5 million in grants to complete reconstruction and rehabilitation of all 1,283 currently certified Rural Critical Access Hospitals. The budgetary impact amounts to $80 million to support the loan guarantees and $5 million for the grants over 10 years. In contrast, since FY2004,

USDA has supported 53 critical access hospitals with $260 million in loans and guarantees.

In FY2007, total loan guarantee budget authority for the entire Community Facilities program amounts to $208 million. Is the staff of the Community Facilities program prepared to handle as many as 1,283 new loan and grant projects?

Will this level of targeted funding for critical access hospitals avoid leaving loan applications for other essential community facilities at a disadvantage?

ENHANCE RURAL INFRASTRUCTURE

In the 2002 farm bill, $360 million was authorized for a backlog of applications for rural development loans and grants. This funding was used exclusively for waste and waste water treatment. *The Administration is proposing $500 million in the 2007 farm bill for backlogged loan and grant applications to further the development of sound infrastructure and "provide the basic services required to ensure a good quality of life or encourage sustainable economic development."*

Will the proposed $500 million eliminate the entire backlog of infrastructure projects? If not, how will the funds be allocated among communities and projects?

The Administration's 2008 budget proposes terminating the Community Facilities Grant program and "Community Connect" Broadband Grants. In the explanation of its farm bill proposals, the Administration notes that the unmet need for the kinds of services provided by the Community Facilities program is substantial. Also, farm bill reauthorization is proposed for broadband access, distance learning, and telemedicine programs. If the next farm bill does indeed reauthorize these programs, what level of funding will the Administration put in future budget requests? Are the farm bill proposal for an additional $500 million for the infrastructure backlog and the FY2008 budget proposal to terminate the programs consistent with one another?

STREAMLINE RURAL DEVELOPMENT PROGRAMS

Loans and grants for business development and expansion are long-standing programs to assist rural areas with economic diversification and new opportunities for rural residents. The programs are targeted to existing rural businesses and startups, public bodies, nonprofit corporations and cooperatives, and they offer assistance in business planning, labor training, and technical assistance. Similarly, loans and grants for infrastructure (e.g., water treatment, technical assistance, broadband development) are also major foci of USDA Rural Development. The Administration's farm bill proposes creating a new "Business Grants Platform," a new "Community Programs Platform," and a "Multi-Departmental Energy Grants Platform" that would consolidate the authorities for many of these programs into single entities.

Multi-Department Energy Grants Platform[5]

The Administration proposes to consolidate USDA energy grant and research program authorities under the Biomass Research and Development Act of 2000. The key Renewable Energy Systems and Energy Efficiency Improvements grant programs would be consolidated under this act with proposed mandatory funding of $500 million over 10 years. In addition, competitive grants under the consolidated authority would be increased to $150 million over 10 years.

In the past, Renewable Energy System funds have assisted a range of renewable energy activities including anaerobic digesters and wind energy systems from across diverse of geographic areas. Will the expanded funding continue to be broadly targeted across different renewable energy types and geographic locations, or will it focus more directly on establishing a viable, self-sustaining cellulosic ethanol industry?

The Department of Energy (DOE) recently announced it would be investing $385 million in six biorefinery projects using cellulosic feedstocks.[6] Is there a need for additional USDA energy grants funding? How will the requested funding in the Administration's energy grants proposal be coordinated with the DOE effort?

What kinds of quality employment and economic development potential for rural America would a multi-department energy grants platform provide?

Business Loan and Loan Guarantee Platform[7]

The Administration proposes to consolidate into existing Business and Industry Loan Program authority several other loan program authorities, prioritize funding for biorefinery construction, and raise the loan guarantee limit on cellulosic plants to $100 million. The Renewable Energy Systems and Energy Efficiency Improvements Loan Guarantee Program would be consolidated under this platform. Proposed increased funding to $210 million would support $2.17 billion of guaranteed loans over 10 years. For cellulosic ethanol projects, the Administration would raise the loan cap to $100 million and eliminate the cap on loan guarantee fees. Finally, the Administration proposes prioritizing funding for the construction of biorefinery projects.

This platform would emphasize energy development in rural areas, particularly cellulosic ethanol production. Although this may be a promising technology, it has yet to be developed commercially, and there remain significant technical obstacles. Based on current technology, and the government's best-educated projections, corn based ethanol will have to account for 34 billion of the Administration's proposed 35 billion gallons of renewable and alternative fuels target for 2017. What is the rationale for the proposed level of funding for such a primitive technology?

What support will be given to other renewable energy technologies (e.g., wind power, solar power)? Is there any concern about crowding out the development of other potentially viable long-run energy solutions by intensifying federal funds on cellulosic ethanol?

While building cellulosic ethanol facilities over the next five years will create some local construction employment, how likely are cellulosic facilities to create new rural competitive advantage for the long term?

Business Grants Platform

The Administration proposes to consolidate the separate legal authorities for five rural grants programs into a single law.

How would this proposed streamlining effort enhance assistance to rural areas? What current obstacles exist within the USDA Rural Development's mission agencies that impede efficient and effective business assistance to rural areas?

Title VI: Rural Development

The Administration's 2008 budget request calls for terminating two of the programs that might have been included in the business grants platform (e.g., Rural Business Enterprise Grants, Rural Business Opportunity Grants). These programs target smaller rural businesses and are important sources of funding for entrepreneurial business activities in rural areas. What is the rationale for eliminating these grants? Will their termination limit the capacity to support more entrepreneurial efforts in rural areas? How would elimination of these programs enhance the efficiency and effectiveness of the remaining programs proposed for consolidation?

Many rural development programs were created in part because rural areas tended to be underserved by the economic development programs administered by other federal agencies (e.g., Department of Commerce, Department of Housing and Urban Development). The Administration's FY2008 budget request considers the Rural Business Enterprise Grants Program and Rural Business Opportunity Grants Program as duplicating programs administered by other federal agencies and proposes their termination. What assurances can be given that rural areas will not be neglected by these other federal agencies? Will funding for similar programs administered by these other federal agencies be increased, or not, to target rural areas for economic development assistance?

Community Programs Platform

This platform would consolidate authorities for water and waste water loans, guarantees, and grants into a single entity. Assorted supplemental authorities would also be consolidated under this platform.

The proposed community programs platform consolidates approximately nine community programs, one of which targets rural areas with high unemployment and/or significant outmigration. Yet, the FY2008 budget proposes terminating two programs, the Economic Impact Grants and Community Facility Grants, stating that they are duplicative of programs in other federal agencies. Will funding be increased in these other federal agencies to target rural areas with high unemployment and/or out-migration? Will essential rural community facilities assisted by the Community Facility Grant Program be supported by other federal agencies?

In: The USDA 2007 Farm Bill Proposal
Editors: J. Womach et al.
ISBN: 978-1-60456-813-4
© 2008 Nova Science Publishers, Inc.

Chapter 7

TITLE VII: RESEARCH

The 2002 farm bill reauthorized ongoing USDA programs in agricultural research, education, extension, and agricultural economics through FY2007, and extended reforms in this mission area that were enacted in 1998 as part of the Agricultural Research, Extension, and Education Reform Act (P.L. 105-185). The agencies that comprise USDA's Research, Extension, and Economics (REE) mission area are the Agricultural Research Service (ARS), the Cooperative State Research, Education, and Extension Service (CSREES), the Economic Research Service (ERS), and the National Agricultural Statistics Service (NASS). ARS is USDA's intramural research agency, comprising more than 100 laboratories nationwide. CSREES distributes annual appropriations to support extramural agricultural research and extension at the land grant colleges of agriculture in the states and U.S. territories. ERS conducts economic research, and NASS is the primary USDA statistical agency.

RESEARCH, EDUCATION AND ECONOMICS (REE) MISSION AREA REORGANIZATION

The Administration proposes the consolidation of ARS and CSREES into a single agency to be called the Research, Education, and Extension Service (REES). The head of the new agency would hold the title of Chief Scientist. The current Research, Extension, and Education mission area would be renamed the Office of Science, with leadership to continue through the Under Secretary and Deputy Undersecretary.

The land grant universities also have put forward a proposal to reorganize USDA's research mission area. Their proposal (referred to as "CREATE-21") would combine ARS and CSREES into one agency, would keep ERS and NASS in the research mission area, and would bring the research function of the Forest Service under the same administrative umbrella as ARS and CSREES research. In addition, CREATE-21 proposes the establishment of a national institute for research on food and agriculture that would support both intra- and extramural science through competitively awarded grants.

In the mid-1970s, the Carter Administration merged ARS, the Cooperative State Research Service (CSRS), and the Extension Service into what was then called the Science and Education Administration (SEA). The same rationales in favor of such a move were cited then as now: that there was costly redundancy at the administrative level, that combining intramural and extramural research programs would result in better coordination, and that more resources should go directly into performing research. SEA was separated back into its three distinct agencies at the beginning of the Reagan Administration. Although the 1994 USDA reorganization combined CSRS and the Extension Service to form CSREES (and brought ERS and NASS into the research mission area), the intramural and extramural research programs have remained separate for more than 25 years. It is widely held that the merger in the 1970s never functioned as intended because little attempt was made to work within and between the previously separate agencies to create a new, combined structure and culture.

What are some of the steps anticipated for the proposed Chief Scientist to take in order to create a well-coordinated single agency that is united behind its mission?

What is the Administration's proposal concerning ERS and NASS? Would they remain under the proposed Office of Science? If not, where would they be placed?

The Forest Service receives roughly $250 million annually through the Department of Interior budget to conduct research related to public and private forest lands. The laboratories where this research is conducted are largely located at land grant institutions, which also receive funds for forestry research through CSREES. The land grant universities' CREATE-21 document proposes bringing Forest Service research under the same administrative umbrella as ARS and CSREES.

What are the Department's reasons for keeping Forest Service research separate in its 2007 farm bill proposal?

ARS receives direct funding through the annual USDA appropriations acts. The states receive federal funds, administered by CSREES, through a variety of block grants (or formula funds) and competitive grants. For the past several years, the Administration has proposed in its annual budget request to cut formula funds to states, while the proposed ARS appropriation remains the same or increases.

Under the proposed merger of ARS and CSREES, how would these different funding mechanisms be treated?
What changes, if any, is the Administration considering to the decades-old formula funding mechanism?
Is the intent of the Administration's proposal to increase the amount of research funding that would be distributed through competitive grants, and decrease the amount distributed through the formula programs? If so, why?

The land grant universities' CREATE-21 proposal contains specific suggestions for how a new, combined funding system could be managed. The stated intent of that proposal is to not disadvantage either ARS or the state research institutions financially as the agencies merge. Key to the CREATE-21 proposal is doubling the current amount of funding for agricultural research and extension over the next seven years.

Does the Administration foresee using mandatory funds to support a significant increase in the total amount of funding available for agricultural research and extension?
How would such funding fit into the Administration's larger proposal for farm program reforms?

AGRICULTURAL BIOENERGY AND BIOBASED PRODUCTS RESEARCH INITIATIVE[8]

The Administration proposes to create an Agricultural Bioenergy and Biobased Products Research Initiative with mandatory annual funding of $50 million for 10 years. The initiative would use existing Agriculture Research Service facilities and scientists and provide competitive grants to universities. These USDA-funded activities would be coordinated with Department of Energy activities. The objectives would be to make agricultural biomass a viable alternative to petroleum and to develop industrial products from the byproducts of bioenergy production.

Will current and pending ARS work be displaced when facilities and scientists shift to the high priority bioenergy topics?

Will the Department of Energy have management control over any of this research funding?

Does the proposal envision any collaboration between public and private research in this area?

SPECIALTY CROP RESEARCH INITIATIVE

The Administration proposes to create a new Specialty Crops Research Initiative with annual mandatory funding of $100 million.

The initiative is said to include both intramural (ARS) and extramural (CSREES) programs. How would the Administration propose to divide the funding between these two categories. With regard to intramural research, how many new scientists might be added to ARS, or would the current staff shift priorities to specialty crops and away from current activities?

There has been a history of mandated research programs going unfunded. What reassurance can USDA give the specialty crop producers that this new initiative will be implemented?

FOREIGN ANIMAL DISEASE RESEARCH

The premier U.S. facility for research on foreign animal diseases is the Plum Island Animal Disease Center, located on an island off the northeastern tip of Long Island, NY. The property of Plum Island was transferred from USDA to DHS in the Homeland Security Act of 2002, although personnel from both USDA's Agricultural Research Service (ARS) and Animal and Plant Health Inspection Service (APHIS) still conduct research there alongside DHS personnel. Many experts agree that the 50-year old Plum Island facility, built in the 1950s, is nearing the end of its useful life and unable to provide the necessary capacity for current biosecurity research.

The Department of Homeland Security is proceeding with plans to replace the aging Plum Island Animal Disease Center with a new "National Bio and Agro-Defense Facility" (NBAF) for research on high consequence foreign animal diseases. Congress already has appropriated $46 million over FY2006-FY2007 for planning and site selection, and the estimated design and construction cost is $451 million. DHS has begun the design process, and already has reviewed submissions from universities and other locations interested in hosting the new facility. In August 2006, it selected a long list of 18 sites in 11 states for further consideration. A final location will be chosen early in 2008, and the current time line calls for construction to be completed in 2013.

Plum Island is the only facility in the United States that is currently approved to study high-consequence foreign livestock diseases, such as foot-and-mouth disease (FMD), because its laboratory has been equipped with a specially designed BSL-3 bio-containment area for large animals that meets specific safety measures. The U.S. Code stipulates that live FMD virus may be used only at coastal islands such as Plum Island, unless the Secretary of Agriculture specifically authorizes the use of the virus on the U.S. mainland (21U.S.C. 113a). *The Administration proposes to change the law to allow research and diagnostics for highly infectious foreign animal diseases on mainland locations in the United States.*

The Plum Island Animal Disease Center and the USDA National Veterinary Services Laboratories (NVSL) in Ames, IA, are the only BSL-3 agriculture facilities in the United States. The United States has no BSL-4 agriculture facilities (the highest biosecurity level); such facilities are located in Canada and Australia. The intended NBAF is likely to be another BSL-3 facility, although a BSL-4 facility has not necessarily been ruled out.

DHS is understood to be already proceeding to build this new laboratory prior to any change the law about FMD research on the mainland. If Congress does not change the law and DHS builds the facility on the mainland, will the Secretary of Agriculture use his regulatory authority to allow such research so that the presumed new facility can be used? Which action should come first, statutory authority or building the facility?

Was USDA consulted about the DHS decision to build a new lab? Does USDA have a preference for location, relative to Plum Island and the USDA personnel who work there?

Does USDA have a seat on the DHS site selection committee?

Critics are concerned that locating the facility in regions where cattle or other livestock are raised may pose an unnecessary risk if security features are breached by terrorism, which is an unpredictable risk compared to accidental or unintentional risks. GAO found security concerns at Plum Island a few years ago. What is the advantage of building such a facility in Kansas, for example, where the consequences of a biosecurity breach could be much more devastating to domestic cattle production than if the facility remained at a coastal site such as Plum Island? How do these risk factors enter the cost-benefit analysis of site selection?

How do the risks compare between animal and human diseases, regarding operating the Centers for Disease Control (CDC) BSL-4 lab in Athens, Georgia, a mainland location, compared to the Plum Island location for agriculture? Which diseases are more likely to spread among a population if released?

In: The USDA 2007 Farm Bill Proposal ISBN: 978-1-60456-813-4
Editors: J. Womach et al. © 2008 Nova Science Publishers, Inc.

Chapter 8

TITLE VIII: FORESTRY

The USDA's Forest Service manages the National Forest System, funds and conducts forestry research, and provides forestry assistance. Most federal forestry programs are permanently authorized. Past farm bills have generally addressed cooperative assistance programs administered by the Forest Service's State and Private Forestry (S and PF) branch.

The Administration's 2007 farm bill proposes four new programs: (1) comprehensive statewide forest planning; (2) competitive landscape-scale forestry grants; (3) a 10-year, $150 million forest wood-to-energy technology development program; and (4) financial and technical assistance to communities for acquiring, planning for, and conserving community forests. The Administration has not proposed reauthorizing the Forest Land Enhancement Program (FLEP). FLEP (a combination of two previously existing landowner assistance programs) was enacted in the 2002 farm bill with mandatory funding of $100 million over the six-year life of the law. Subsequently, at the request of the Administration, funding authority was reduced to $49.5 million.

COMPREHENSIVE STATEWIDE PLANNING

The Administration is proposing a new program of financial and technical assistance to state forestry agencies to develop and implement Statewide Forest Resource Assessments and Plans to address the increasing public demand for forest products and amenities, pressure on landowners to convert forests to other uses, and risk from wildfire.

Would the proposed statewide planning, technology development, and community forests be more effective at providing for the growth in demand for forest products and amenities than a direct landowner assistance program, such as FLEP?

Does the lack of private landowner assistance in the 2007 proposal constitute a conclusion that the programs have been ineffective? How many private landowners have been assisted annually over the past decade by the existing cost-share assistance programs, and what are the results of these efforts?

How can national direction for statewide forest planning best provide the flexibility to address such diverse forests as those in Iowa and those in Florida? Are the various state forestry organizations unable or unwilling to undertake statewide forest planning without federal direction and oversight? How is this new planning effort to be funded, given the Administration proposal to cut FY2008 forest stewardship funding (for financial and technical assistance to states) by 41%? How would statewide forestry planning address the identified growth in demand for forest products and amenities and in low-value biomass that degrades forests and increases wildfire risk?

LANDSCAPE-SCALE COMPETITIVE GRANT PROGRAM

The Administration's farm bill proposal includes a new landscape-scale forestry competitive grant program "to develop innovative solutions that address local forest management issues; develop local nontraditional forest product markets; and stimulate local economies through creation of value-added forest product industries." The Administration identifies as significant problems the aging of family forest landowners and the potential fragmentation of forests over the next two decades.

How would "landscapes" be defined for the grants? Would competitive landscape-scale grants require cooperative involvement of all or most landowners within the landscape? If the grants are to foster nontraditional markets and value-added industries, would they even be related to the landscape and the landowners?

The proposal states that the program "would ensure a comprehensive, coordinated approach to forest management and would ensure collaboration across ownership and jurisdictional boundaries." What proportion of the landowners or of the lands need to be involved for a landscape to be eligible for a grant? How can landowners, including the federal government, be enocuraged to cooperate? How would the landscape grant proposals be assessed and compared; that is, what criteria would be used to make the grants competitive? Does the Forest Service have the needed expertise to implement a competitive landscape-scale grant program? Do landscape-scale grants and community forests move away from private, individual forestland ownership, and promote communal forest ownership?

FOREST WOOD FOR ENERGY[9]

The Administration is proposing a new 10-year, $150 million wood-to-energy program to accelerate development and use of new technologies to use the substantial amounts of low-grade woody biomass that degrade forest health and exacerbate wildfire risks and are of little commercial value.

What are the program goals for this proposal? How will progress and effectiveness be measured?

What is the potential to convert woody biomass to cellulosic ethanol, and how does this compare with the potential to burn woody biomass to produce electricity? What are the costs and the biomass conversion factors for ethanol conversion and for electricity production? What other factors — capital costs, infrastructure, collection and hauling opportunities, etc. — might be critical for improved utilization of low-value woody biomass for energy? Might any of these factors be more limiting than technology development and deployment? What programs exist to address these other factors?

COMMUNITY FORESTS WORKING LANDS PROGRAM

The Administration's 2007 farm bill proposes a community forests working lands program to provide communities with the financial assistance to acquire and conserve forests and the technical assistance to plan for the use and conservation of those forests. This program would particularly address the problem of producing goods and services from forest at the urban fringe.

How does the proposed community forests program differ from the existing Forest Legacy program?

What is the federal role and federal responsibility in funding and otherwise assisting communities in acquiring and conserving local forestlands?

In: The USDA 2007 Farm Bill Proposal ISBN: 978-1-60456-813-4
Editors: J. Womach et al. © 2008 Nova Science Publishers, Inc.

Chapter 9

TITLE IX: ENERGY

Title IX of the 2002 farm bill (P.L. 107-171) represented the first-ever energy title in a farm bill and included nine provisions addressing agriculture-based renewable energy systems. *USDA's proposed 2007 farm bill outlines modifications to programs that expand federal research on renewable fuels and bioenergy; and re-authorizes, revises, and expands programs intended to provide assistance for the advancement of renewable energy production and commercialization.* However, several expiring provisions from the 2002 farm bill are not mentioned. These include the Biorefinery grants (Section 9003), the Biodiesel Fuel Education Program (Section 9004), the Energy Audit and Renewable Energy Development Program (Section 9005), the Memorandum of Understanding between the Secretary of Agriculture and the Secretary of Energy concerning hydrogen and fuel cell technologies (Section 9007), and Cooperative Research and Development on Carbon Sequestration (Section 9009). It is also noteworthy that several of these same provisions went unfunded during the life of the 2002 farm bill.

Is it USDA's intention that expiring provisions in the 2002 farm bill be dropped from future legislation? These provisions were never funded or implemented during the past five years. Would the USDA support funding these expiring provisions if they are reauthorized by Congress?

What progress has been made to improve coordination between USDA and the Department of Energy? Is there still room for major improvements or are the two departments already fairly efficient in coordinating energy development activities?

CELLULOSIC BIOENERGY PROGRAM

USDA's energy proposal calls for a substantial increase in funding under the loan guarantee and grants program of the Renewable Energy Systems and Energy Efficiency Improvements program (otherwise referred to as the Renewable Energy Program). In addition, these programs are to be managed to provide preference to projects that focus on cellulosic ethanol.

What accomplishments can be claimed by the Renewable Energy Program in furthering the development of renewable energy in general and biofuels in particular?

Current thinking is that, once the technology is developed, cellulosic ethanol will expand rapidly to take advantage of cheap feedstocks, such as switchgrass, that can be produced on marginal lands. *The farm bill proposal includes some incentives to encourage development of a cellulosic-based ethanol industry.*

However, there are still many questions surrounding the potential of cellulosic ethanol and the likely economic implications associated with a major expansion of cellulosic ethanol production.

Biomass material is bulky and poses serious challenges for harvesting, transportation, and storage. How much of USDA's research funding would be targeted to these types of issues?

If cellulosic feedstocks are produced on marginal lands, would they compete directly with cattle forage?

Many conservation and wildlife proponents are concerned about the possibility of degrading Conservation Reserve Program (CRP) acreage for cellulosic feedstock production. What assurances might be offered in this regard?

If the cellulosic ethanol industry takes off, will there still be room for the corn-based ethanol industry? Would a cellulose-based ethanol industry shift its geographic location towards the cheaper lands and feedstocks of the prairies and forests of America, leaving behind corn-based plants of the Corn Belt? If some version of this were to develop, what would be the outlook for corn-based ethanol plants? What would happen to those individuals, many from small towns across America, that have poured their savings into ethanol plants?

American ethanol blenders receive at least partial protection from foreign competition by a $0.54 per gallon tariff on imported ethanol. Although the Caribbean Basin Initiative allows for modest entry of ethanol from several Caribbean countries, the tariff clearly works against ethanol from Brazil. This tariff was recently extended through 2008 (by P.L. 109-432). Some ethanol supporters argue that this tariff prevents the development of a national distribution network by limiting access to adequate ethanol supplies by ethanol blenders in the major coastal regions of the United States such as New York, Florida, and California.

Does the tariff on ethanol imports create a supply problem for major metropolitan areas distant from the Corn Belt? If the import tariff can be justified as providing essential protection for the ethanol industry, why is there no similar tariff on either biodiesel or palm oil to protect the more nascent U.S. biodiesel industry?

In addition to import protection, the U.S. ethanol sector receives substantial support from (1) a tax credit of $0.51 to fuel blenders for every gallon of ethanol blended with gasoline, and (2) a Renewable Fuels Standard (RFS) that mandates a renewable fuels blending requirement for fuel suppliers that grows annually from 4 billion gallons in 2006 to 7.5 billion gallons in 2012. A recent survey of both federal and state subsidies in support of ethanol production reported that total annual federal support is somewhere in the range of $5.1 to $6.8 billion.[10] USDA's energy proposal continues the trend of strong support to the biofuels sector.

Is there concern that these subsidies for a single technology, in this case the combustion engine and biofuels, may deter or limit the development of new or as-yet unknown future technologies that might otherwise provide more sustainable long-run solutions to the United States' energy situation?

EXPAND OF BIOBASED PRODUCTS MARKETS

The Administration recommends that the law authorizing the Federal Procurement of Biobased Products program (section 9002 of the 2002 farm bill) be changed to improve the effectiveness and administration of the program. Also, additional mandatory funding of $2 million per year is recommended.

Federal law mandates the use of a sizeable amount of renewable fuel and it appears future growth will not need the help of federal agency procurement. If renewable fuel is not the focus of the federal procurement program, what will be the focus?

CONSOLIDATE ENERGY BUSINESS LOAN AUTHORITIES UNDER THE BIOMASS RESEARCH AND DEVELOPMENT ACT[11]

The Administration proposes to consolidate into existing Business and Industry Loan Program authority several other loan program authorities, prioritize funding for biorefinery construction, and raise the loan guarantee limit on cellulosic plants to $100 million. The Renewable Energy Systems and Energy Efficiency Improvements Loan Guarantee Program would be consolidated under this platform. Proposed increased funding to $210 million would support $2.17 billion of guaranteed loans over 10 years. For cellulosic ethanol projects, the Administration would raise the loan cap to $100 million and eliminate the cap on loan guarantee fees. Finally, the Administration proposes prioritizing funding for the construction of biorefinery projects.

This platform would emphasize energy development in rural areas, particularly cellulosic ethanol production. Although this may be a promising technology, it has yet to be developed commercially, and there remain significant technical obstacles. Based on current technology, and the government's best-educated projections, corn based ethanol will have to account for 34 billion of the Administration's proposed 35 billion gallons of renewable and alternative fuels target for 2017. What is the rationale for the proposed level of funding for such a primitive technology?

What support will be given to other renewable energy technologies (e.g., wind power, solar power)? Is there any concern about crowding out the development of other potentially viable long-run energy solutions by intensifying federal funds so narrowly on cellulosic ethanol?

While building cellulosic ethanol facilities over the next five years will create some local construction employment, how likely are cellulosic facilities to create new rural competitive advantage for the long term?

CREATE A MULTI-DEPARTMENT ENERGY GRANTS PROGRAM[12]

The Administration proposes to consolidate USDA energy grant and research program authorities under the Biomass Research and Development Act of 2000. The key Renewable Energy Systems and Energy Efficiency Improvements grant programs would be consolidated under this act with proposed mandatory funding of $500 million over 10 years. In addition, competitive grants under the consolidated authority would be increased to $150 million over 10 years.

In the past, Renewable Energy System funds have assisted a range of renewable energy activities including anaerobic digesters and wind energy systems from across diverse of geographic areas. Will the expanded funding continue to be broadly targeted across different renewable energy types and geographic locations, or will it focus more directly on establishing a viable, self-sustaining cellulosic ethanol industry?

The DOE recently announced it would be investing $385 million in six biorefinery projects using cellulosic feedstocks.[13] Is there a need for additional USDA energy grants funding? How will the requested funding in the Administration's energy grants proposal be coordinated with the DOE effort?

What kinds of quality employment and economic development potential for rural America would a multi-department energy grants platform provide?

CRP BIOMASS RESERVE[14]

The Conservation Reserve Program (CRP) and Conservation Reserve Enhancement Program (CREP) remove active cropland into conservation uses, typically for 10 years, and provide annual rental payments based on the agricultural rental value of the land and cost-share assistance. Conversion of the land must yield adequate levels of environmental improvement to qualify (environmental benefits index). CRP is the largest land retirement program with spending of $1.828 billion in FY2005. The total program acreage is limited to 39.2 million.

The Secretary is recommending reauthorization of this program with an enhanced focus on lands that provide the most benefit for environmentally sensitive lands. Priority would be given to whole-field enrollment for lands utilized for energy-related biomass production. Biomass would be harvested after nesting season and rental payments would be limited to income foregone or costs incurred by the participant to meet conservation requirements in those years biomass was harvested for energy production.

The proposal may appear to some to have two conflicting components with regard to CRP. If it is desirable to focus CRP on multi-year idling of more environmentally sensitive lands, what is the rationale for proposing the harvesting of biomass on those lands? Could this harvesting conflict with the purpose of the program?

If it is decided that high demand for commodities dictates that less land should be in the CRP, how would priorities be set for land to be released?

MANDATORY FUNDING FOR COMPETITIVE GRANTS UNDER THE BIOMASS RESEARCH AND DEVELOPMENT ACT[15]

The Administration proposes to consolidate USDA energy grant and research program authorities under the Biomass Research and Development Act of 2000.

The key Renewable Energy Systems and Energy Efficiency Improvements grant programs would be consolidated under this act with proposed mandatory funding of $500 million over 10 years. *In addition, competitive grants under the consolidated authority would be increased to $150 million over 10 years.*

See the questions under the previous heading in this chapter titled "Create a Multi-Department Energy Grants Program."

MANDATORY FUNDING FOR USDA/UNIVERSITY COLLABORATIVE RESEARCH[16]

The Administration proposes to create an Agricultural Bioenergy and Biobased Products Research Initiative with mandatory annual funding of $50 million for 10 years. The initiative would use existing Agriculture Research Service facilities and scientists and provide competitive grants to universities. These USDA-funded activities would be coordinated with Department of Energy activities. The objectives would be to make agricultural biomass a viable alternative to petroleum and to develop industrial products from the byproducts of bioenergy production.

Will current and pending ARS work be displaced when facilities and scientists shift to the high priority bioenergy topics?

Will the Department of Energy have management control over any of this research funding?

Does the proposal envision any collaboration between public and private research in this area?

FOREST WOOD FOR ENERGY[17]

The Administration is proposing a new 10-year, $150 million wood-to-energy program to accelerate development and use of new technologies to use the substantial amounts of low-grade woody biomass that degrade forest health and exacerbate wildfire risks and are of little commercial value.

What are the program goals for this proposal?

What is the potential to convert woody biomass to cellulosic ethanol, and how does this compare with the potential to burn woody biomass to produce electricity? What are the costs and the biomass conversion factors for ethanol conversion and for electricity production? What other factors — capital costs, infrastructure, collection and hauling opportunities, etc. — might be critical for improved utilization of low-value woody biomass for energy? Might any of these factors be more limiting than technology development and deployment? What programs exist to address these other factors?

Chapter 10

TITLE X: MISCELLANEOUS

FEDERAL CROP INSURANCE

The federal crop insurance program is permanently authorized so it does not require renewal in the 2007 farm bill. Major enhancements to the program have been authorized in legislation on several occasions since 1980 (usually outside of the farm bill process). Most recently, the Agriculture Risk Protection Act of 2000 (P.L. 106-224) put $8.2 billion in new federal spending measures over a five-year period into the program primarily through more generous premium subsidies to help make the program more affordable to farmers and increase farmer participation. Since 2000, the federal subsidy to the crop insurance program has averaged about $3.3 billion per year.

Although the scope of crop insurance has widened significantly over the past 25 years and premium subsidies have increased, the stated goal of eliminating disaster payments has not been achieved. Until the 2005 crop, Congress provided *ad hoc* disaster payments to farmers in virtually every year since 1988 that witnessed substantial weather-related crop losses. The disaster assistance has been made available regardless of whether a producer had an active crop insurance policy.

The Administration's farm bill proposal contains several crop insurance recommendations intended to enhance participation; address issues of waste, fraud and abuse; reduce costs; and reduce the need for emergency supplemental disaster payments. *One of the more significant proposed changes to the program would be to allow participating farmers to purchase insurance for the portion of their production that is part of their deductible, and not currently covered by crop insurance.* Under this supplemental deductible coverage, a producer could purchase an additional policy, and a payment would be made when losses in the producer's county exceed a certain threshold. *The Administration also recommends several cost-saving measures to the program including reducing premium subsidies by 2 to 5 percentage points, charging premiums for the catastrophic level of coverage (which currently is premium-free), and requiring the private insurance companies (which now sell and service the policies) to absorb more of the cost of the program. Finally, farmers would be required to purchase crop insurance as a prerequisite for participating in the farm commodity support programs.*

The estimated annual average cost of the supplemental deductible coverage that the Administration proposes is $35 million. Over the last twenty years, Congress has provided an average of about $2 billion per year in supplemental disaster payments. How would this proposed program preclude the pressure for Congress to enact multi-billion dollar disaster payment programs each year?

What effect would the Administration proposals to reduce the federal cost of the crop insurance program by increasing farmer-paid premiums have on farmer participation in the program?

A 1994 crop insurance act required the purchase of a crop insurance policy as a prerequisite for participating in the farm commodity programs. Farm groups were strongly opposed to this provision and fought successfully to have it eliminated in the 1996 farm bill. What reaction might be expected from farm groups to the current proposal for mandatory linkage?

SECTION 32 FRUIT AND VEGETABLE PURCHASES FOR NUTRITION PROGRAMS

"Section 32" is a permanent appropriation that since 1935 has earmarked the equivalent of 30% of annual customs receipts to support the farm sector through a variety of activities. Today, most of this appropriation (now approximately $7 billion yearly) is transferred to the U.S. Department of Agriculture (USDA) account that funds child nutrition programs. However, a smaller — but still significant — amount of Section 32 money is set aside each year to purchase non-price-supported commodities directly and provide them to schools and other feeding sites. Some of these purchases are "entitlement" commodities that are required to be made under the school

Title X: Miscellaneous

lunch act. Others are "bonus" commodities, acquired through emergency surplus removal activities. The total value of both types of commodities now exceeds $900 million per year. The purchases are made by USDA's Agricultural Marketing Service (AMS). Included within these combined ("mandated" and "bonus") Section 32 totals, fruit and vegetable purchases over the last five years have averaged $308 million per year, according to USDA.

USDA proposes to increase purchases of fruits and vegetables using Section 32 authority by at least $200 million per year, but the farm bill budget indicates no score above the OMB baseline. Why is this proposal not reflected in the Administration's FY2008 budget? In other words, how does the Administration propose to cover the cost of these increased fruit and vegetable purchases?

If new spending would not be created, which activities or food purchases would be reduced to pay for these increases? For example, Section 32 is now also used to purchase animal products including meats, poultry, and seafood. Would the Administration proposal result in fewer purchases of these products? If not, why?

The Department routinely has funds remaining in the Section 32 account at the end of each fiscal year, which are "carried over" into the next fiscal year to be used in Section 32. What is this level of unobligated funds, on average, and does USDA intend to reduce the size of this carryover to pay for new fruit and vegetable purchases? If so, won't that leave even less carryover in future years?

Does this proposal call for any new legislative authority, and if not, how can Congress be assured that the initiative would be carried out by future Administrations?

How does the Department currently determine what proportions of its Section 32 commodity acquisitions go to various domestic nutrition programs, and how would it do so for the proposed increases?

How does this proposal differ from the separate Administration initiative providing for $50 million yearly in other new fruit and vegetable purchases for domestic nutrition programs? How would it be funded?

Why does the Department need, and use, such broad legislative authority to administer Section 32 programs, particularly "bonus" surplus removals?

ORGANIC AGRICULTURE

The Administration's 2007 farm bill proposal recommends considerably more funding for research and marketing programs, to support the continuing growth of the organic farming sector.

How is the Department proposing to provide this new mandatory funding?

USDA's farm bill initiative states that gaps in the organic regulations may need to be addressed in order to better support enforcement activity. But more enforcement would also require more personnel and resources. How would the Department provide funding for the increased program oversight and enforcement that could be necessary as the number of certified operations increases?

In: The USDA 2007 Farm Bill Proposal
Editors: J. Womach et al.

ISBN: 978-1-60456-813-4
© 2008 Nova Science Publishers, Inc.

APPENDIX. ADMINISTRATION'S COST ESTIMATE

Administration's 2007 Farm Bill Proposal Baseline and Estimated Change from Baseline Budget Authority, FY2008-2017 Dollars in Millions

Title and Proposals	Current Services Baseline	Proposed Change from Baseline
Title I — Commodities		
Marketing Assistance Loans	$8,807	-$4,500
Posted County Price and Loan Repayment Changes	na	-250
Direct Payment Program	52,491	5,500
Direct Payment Bonus for Beginning Farmers	na	250
Revenue-based Counter-Cyclical Payment Program	11,245	-3,700
Payment Limits and Eligibilty	na	-1,500
Section 1031 Farmland 1031 Exchanges	na	-30
Dairy	613	793
Sugar Program	1,410	-1,107
Special Cotton Competitiveness Program	na	na
Planting Flexibility Limitations	na	na
Retire Crop Bases in Nonagricultural Use	na	na
Conservation Enhancement Payment Option	na	50
Sodsaver	na	na
Continuing WTO Compliance	na	na
Total	74,566	-4,494
Title II — Conservation		
Revised Environmental Quality Incentives Prog. (EQIP)	13,640	4,250
Regional Water Enhancement	na	1,750
Wildlife Habitat Incentives	0	na

(Continued).

Title and Proposals	Current Services Baseline	Proposed Change from Baseline
Ground and Sruface Water Conservation	600	na
Agri Management Assistance	100	na
Conservation Innovation Grants	200	1,000
Klamath	0	0
Conservation Security Program (CSP)	7,977	500
Private Lands Protection Program	970	900
Grasslands Reserve	0	na
Farm and Ranch Land Protection	970	na
Healthy Forest Reserve	a	na
Conservation Reserve Prog. (CRP)	25,656	0
Wetlands Reserve Program	455	2,125
Conservation Access for Beginning/Limited Resource Producers	na	0
Market-based Conservation	na	50
Merit-based Funding	na	0
Emergency Landscape Restoration Program	a	a
Total	48,698	7,825
Title III — Trade		
Technical Assistance for Specialty Crops (TASC)	0	68
Market Access Program (MAP)	2,000	250
SPS Issues Grant Program	na	20
Support International Trade Standard Setting Activities	na	15
Trade Disputes Technical Assistance	a	0
Trade Capacity Building and Post-Conflict Ag. Extension	na	20
Export Credit Gurarantee Reforms	na	0
Facility Guarantee Program	0	16
Repeal of EEP and Trade Strategy Report	0	0
Cash Authority for Emergency Food Aid	na	0
Total	2,000	389
Title IV — Nutrition		
Food Stamp Program	436,145	-66
Working Poor and Elderly	na	1,378
Streamlining, Modernization and Program Integrity	na	-1,544
Nutrition Education	na	100
The Emergency Food Assistance Program (TEFAP)	1,400	0
(FDPIR)	913	27
Promoting Healthful Diets	na	506
School Lunch - Fruit and Vegetable Purchases	na	500
School Purchase Study	na	6
Senior Farmers' Market Program	150	0
Total	438,608	467

Appendix: Administration's Cost Estimate

Title and Proposals	Current Services Baseline	Proposed Change from Baseline
Title V — Credit		
Loans to Beginning and Socially Disadvantaged Farmers	a	0
Beginning farmer and Rancher Downpayment Loan Program	a	0
FSA Direct Loan Limits	a	0
Total	a	0
Title VI — Rural Development		
Rural Critical Access Hospitals	a	85
Enhancing Rural Infrastructure	a	500
Streamlining Rural Development Programs	na	0
Total	a	585
Title VII — Research		
REE Mission Area Reorganization	na	0
Bio-Energy and Bio-Based Products Research Initiative	na	500
Specialty Crop Research Initiative	na	1,000
Foreign Animal Disease Research on U.S. Mainland	na	0
Total	na	1,500
Title VIII — Forestry		
Comprehesive Statewide Forest Planning	na	0
Landscape Scale Forestry Competitive Grant Program	na	0
Forest Wood to Energy	na	150
Community Forests Working Lands Program	na	0
Total	na	150
Title IX — Energy		
Biomass Research and Development Act Initiative	0	150
Renewable Energy Systems and Energy Efficiency — Grants	a	500
Renewable Energy Systems and Energy Efficiency — Loans	a	210
Commodity Credit Corporation Bioenergy Program	na	100
Federal Biobased Product Procurement Program	0	18
Total	0	978
Title X — Miscellaneous		
Crop Insurance Program	54,641	-2,511
Supplemental Deductible Coverage	na	350
Expected Loss Ratio	na	-1,071
Data Mining Information Sharing	na	0
Program Compliance	na	0
Research and Development	na	0
Renegotiation of Standard Reinsurance Agreement	na	0
Increase Participation While Controlling Costs	na	-1,790
Dairy Research and Promotion Assessment Fairness	na	0
Organic Farming Initiatives	na	61
Increase Section 32 Purchases of Fruits and Vegetables	na	0

(Continued).

Title and Proposals	Current Services Baseline	Proposed Change from Baseline
Total	54,641	-2,450
Grand Total	$618,513	$4,950

Notes: na = not applicable (proposal not in baseline or included in other base or proposed programs); a = discretionary account; 0 = no mandatory spending.

Source: USDA, 2007 Farm Bill Proposals, Washington, DC, pp. 181-183.

REFERENCES

[1] This proposal is repeated in the Energy title of the USDA recommendations and the questions posed here are repeated under the CRP Biomass Reserve heading in the energy section.

[2] The Administration also proposes applying a food stamp "cash-only" categorical eligibility rule to recipients of Supplemental Security Income (SSI) benefits. However, it is unlikely that a "cash-only" food stamp rule would have any effect on SSI recipients because virtually all, if not all, SSI recipients get cash SSI payments (or are authorized to receive them).

[3] States also are eligible for a total of $18 million in bonus payments for high performance in providing program access and processing applications in a timely manner. In FY2005, 13 states received these bonuses.

[4] This Section 32 farm bill recommendation is listed in the USDA report in the Nutrition title and in the Miscellaneous title. The questions posed here are duplicated again in the Miscellaneous title.

[5] This proposal is repeated in the Energy title of the USDA recommendations and the questions posed here are repeated under the heading "Create a Multi-Department Energy Grants Program" in the energy section.

[6] Department of Energy, Office of Public Affairs, *DOE Selects Six Cellulosic Ethanol Plants for Up to $385 Million in Federal Funding*, press release dated February 28, 2007.

[7] This proposal is repeated in the Energy title of the USDA recommendations and the questions posed here are repeated under the heading "Consolidate Energy Business Loan Authorities" under the Biomass Research and Development Act in the energy section.

[8] This proposal is repeated in the Energy title of the USDA recommendations and the questions posed here are repeated under the "Mandatory Funding for USDA/University Collaborative Research" in the energy section.

[9] This proposal is repeated in the Energy title of the "USDA recommendations" and the questions posed here are repeated under the "Forest Wood for Energy" heading in the forestry section.

[10] Doug Koplow, *Biofuels — At What Cost? Government Support for Ethanol and Biodiesel in the United States*, Global Subsidies Initiative of the International Institute for Sustainable Development, Geneva, Switzerland, October 2006; available at [http://www.globalsubsidies.org].

[11] This proposal and the questions are repeated from the Rural Development title.

[12] This proposal and the questions are repeated from the Rural Development title.

[13] Department of Energy, Office of Public Affairs, *DOE Selects Six Cellulosic Ethanol Plants for Up to $385 Million in Federal Funding*, press release dated February 28, 2007.

[14] This biomass reserve recommendation also is listed in the "Conservation" title as part of the "Conservation Reserve Program" recommendation and this entry is a duplicate of the questions posed in that section.

[15] This proposal and the questions are repeated from the Rural Development title.

[16] This biomass reserve recommendation also is listed in the "Research" title as the "Agriculture Bioenergy and Biobased Products Research Initiative" and this entry is a duplicate of the questions posed in that section.

[17] This forest wood recommendation also is listed in the "Forestry" title and this entry is a duplicate of the questions posed in that section.

INDEX

A

access, 10, 59, 60, 77, 91
accessibility, 38
accidental, 70
accuracy, 46
acquisitions, 54, 55, 85
acute, 59
ad hoc, 5, 83
adaptation, 16
adjusted gross income, 6
adjustment, 11
administration, 42, 46, 47, 48, 49, 77
administrative, 13, 15, 17, 18, 19, 21, 31, 32, 42, 43, 44, 47, 48, 50, 51, 66
Afghanistan, 29
age, 4, 38
AGI, 6, 7
aging, 69, 72
agricultural, vii, 15, 16, 17, 19, 20, 25, 26, 27, 28, 29, 30, 31, 32, 33, 65, 67, 68, 79, 81
agricultural commodities, 20, 33
agricultural exports, 25, 26, 27, 30, 31, 32
agricultural sector, 27
agriculture, 6, 7, 12, 26, 29, 30, 40, 65, 66, 69, 70, 75
aid, 32, 33, 34, 37, 43

alternative, 6, 9, 10, 21, 48, 62, 68, 78, 81
AMS, 55, 85
anaerobic, 61, 79
anaerobic digesters, 61, 79
analysts, vii
Animal and Plant Health Inspection Service (APHIS), 68
animal diseases, 68, 69
animal health, 27
animals, 69
antimicrobial, 27
application, 17
appropriations, 25, 26, 27, 33, 39, 52, 54, 57, 65, 67
argument, 7
ARS, 65, 66, 67, 68, 81
ash, 91
assets, 35, 36, 38, 39, 43, 44
associations, 22, 26
Athens, 70
attention, 21
Australia, 69
authority, viii, 14, 17, 25, 26, 27, 28, 29, 31, 33, 34, 42, 44, 45, 54, 55, 56, 57, 58, 60, 61, 62, 69, 71, 78, 79, 80, 85
availability, 33, 34, 57
averaging, 18

B

Balanced Budget Act, 59
barrier, 50
barriers, 5, 10, 25, 26, 27, 28
basic services, 60
beef, 27
beet sugar, 10
benefits, 2, 5, 7, 12, 13, 19, 36, 37, 38, 39, 40, 41, 42, 43, 44, 45, 46, 47, 53, 57, 79, 91
bilateral, 10
biodiesel, 77
biofuels, 76, 77
biomass, 19, 68, 72, 73, 80, 81, 92
biorefinery, 61, 62, 78, 79
biosecurity, 68, 69, 70
biotechnology, 27
block grants, 67
bonus, 46, 55, 56, 85, 91
borrowing, 25, 26, 27, 29
Boston, 8
Brazil, 77
broadband, 60, 61
Broadband, 60
BSE, 27
burn, 73, 81
business, 5, 61, 62, 63
butter, 8

C

California, 16, 77
Canada, 69
cane sugar, 10
capacity, 17, 29, 30, 63, 68, 88
capacity building, 30
capital, 7, 73, 81
capital cost, 73, 81
capital gains, 7
carbon, 75
Caribbean, 77
Caribbean Basin Initiative, 77
Caribbean countries, 77
cash aid, 43
cattle, 69, 76
CBO, 16
cellulose, 19, 76
cellulosic, 61, 62, 73, 76, 78, 79, 81
cellulosic ethanol, 61, 62, 73, 76, 78, 79, 81
Centers for Disease Control (CDC), CCC, 1, 10, 25, 26, 27, 29, 30, 33, 70
certainty, 5, 13
certificate, 2, 6
certification, 22
child nutrition programs, 33, 55, 84
children, 43
circulation, 41
citizens, 23
civilian, 40
classes, 57
classified, 20
collaboration, 68, 72, 81
collateral, 6
colleges, 65
Colorado, 42
combat, 40
combustion, 77
commercial, 7, 22, 30, 58, 73, 81
commercial bank, 58
commercialization, 75
commodities, 1, 2, 3, 4, 7, 11, 12, 19, 20, 26, 27, 30, 31, 32, 33, 34, 40, 49, 53, 54, 55, 80, 84
commodity, vii, 1, 2, 3, 4, 5, 6, 7, 8, 9, 12, 13, 14, 18, 21, 28, 31, 32, 34, 54, 55, 84, 85
Commodity Credit Corporation (CCC), 1, 15, 25, 30, 89
communication, 49
communities, 60, 71, 73, 74
community, 59, 60, 63, 71, 72, 73, 74
compensation, 28
competition, 12, 27, 29, 51, 77
competitive advantage, 62, 78
compliance, 20, 21

components, 12, 17, 18, 19, 80
computer, 42
computers, 42
conflict, 19, 28, 80
conformity, 31
confusion, 15, 23, 51
Congress, vii, 1, 6, 11, 16, 22, 32, 33, 55, 69, 75, 83, 84, 85
Congressional Budget Office (CBO), 1
Congressional Research Service (CRS), vii, 1
conservation, vii, 13, 15, 16, 17, 18, 19, 20, 21, 22, 73, 76, 79, 80
Conservation Security Program, 13, 17, 88
consolidation, 6, 15, 17, 18, 20, 23, 63, 65
construction, 62, 69, 78
consumption, 8, 16
continuing, 18, 86
contracts, 7, 13, 18, 50, 51
control, 10, 37, 50, 57, 68, 81
control group, 37
conversion, 13, 19, 21, 73, 81
coordination, 66, 75
corn, 2, 3, 6, 11, 62, 76, 78
corporations, 61
cost saving, 48
cost-benefit analysis, 70
cost-effective, 36, 39
costs, 1, 4, 11, 13, 19, 21, 31, 32, 36, 37, 38, 39, 40, 41, 42, 43, 44, 47, 51, 58, 73, 80, 81, 83
cost-sharing, 16
cotton, 1, 2, 3, 4, 6, 7, 11, 12
counter-cyclical payments, 5, 6, 9, 11, 13
counter-cyclical program, 4, 6, 9
coverage, 84
covering, viii
cows, 9
credit, vii, 1, 22, 30, 31, 57
crop disaster, 4
crop insurance, vii, 4, 5, 83, 84
crop production, 20
crops, 1, 2, 5, 6, 10, 18, 26
crowding out, 62, 78
CRP, 15, 19, 20, 76, 79, 80, 88, 91
culture, 66
current limit, 58

D

dairy, 8, 9
dating, 58
debt, 30
decision making, 28
decisions, 12, 43
deductible, 83, 84
deduction, 37
deficiency, 2, 6, 11
definition, 17, 23
degrading, 76
degree, 31
delivery, 58, 59
demand, 19, 31, 71, 72, 80
Department of Commerce, 63
Department of Defense, 54
Department of Energy (DOE), 61, 68, 75, 79, 80, 81, 91, 92
Department of Health and Human Services, 49
Department of Homeland Security (DHS), 68, 69
Department of Housing and Urban Development, 63
Department of Interior, 66
developed countries, 27
developing countries, 34
development assistance, 30, 63
diets, 49, 55
disabled, 35, 36, 38, 39
disaster, 4, 5, 6, 22, 83, 84
disaster assistance, 4, 5, 6, 22, 83
discretionary, 22, 23, 29, 57, 90
diseases, 70
dispute settlement, 27, 29
disputes, 25, 28, 29
distance learning, 60

distortions, 2, 7
distribution, 30, 33, 51, 53, 57, 77
diversification, 16, 61
Doha, 9, 14, 32
drought, 22
dry, 1, 2
duplication, 17
duration, 13

E

earnings, 37
economic, 4, 5, 7, 8, 53, 60, 61, 63, 65, 76, 79
economic development, 60, 61, 63, 79
Economic Research Service, 65
economics, 65
economies, 20, 33, 72
economy, 7
education, 32, 33, 35, 37, 38, 39, 48, 49, 65, 66, 75, 88
elderly, 35, 36, 38, 39
electricity, 73, 81
electronic, 41
eligibility standards, 49
eligible countries, 31
ELS, 11
emerging markets, 30
employees, 40, 44
employment, 30, 37, 61, 62, 78, 79
endangered, 18
energy, vii, 19, 61, 62, 71, 73, 75, 76, 77, 78, 79, 80, 81, 91, 92
enhancement, 16, 17, 18, 19, 20, 31, 71, 79, 87
enrollment, 19, 20, 79
entrepreneurial, 63
environmental, 13, 19, 22, 79
Environmental Quality Incentives Program (EQIP), 15, 16, 17, 21, 87
equity, 6, 13
erosion, 20
ERS, 65, 66
ethanol, 62, 73, 76, 77, 78, 81

Ethanol, 91, 92
Eurocentric, 28
European, 31
European Union, 31
evidence, 26, 27, 31
expenditures, 1, 4, 9, 14
expertise, 73
experts, 27, 68
export competitiveness, 11
export credit guarantees, 25
export subsidies, 11, 25, 31, 32
exports, 4, 31

F

failure, 50
faith, 50
family, 72
FAO, 28
farm, vii, 1, 2, 3, 4, 5, 6, 7, 8, 9, 10, 11, 12, 13, 14, 15, 16, 17, 20, 21, 25, 26, 27, 29, 30, 31, 32, 33, 34, 44, 47, 51, 54, 55, 57, 58, 60, 61, 65, 67, 71, 72, 73, 75, 76, 77, 83, 84, 85, 86, 91
Farm Bill, i, iii, vii, 87, 90
farm bills, 15, 71
Farm Credit System (FCS), 58
Farm Service Agency (FSA), 22, 57, 58, 89
farmers, 2, 3, 4, 5, 6, 7, 8, 10, 11, 13, 18, 20, 21, 22, 49, 52, 57, 83
farming, 4, 19, 86
farmland, 5, 7, 8, 22
Farmland Protection Program, 18
farms, 6, 7, 8, 9, 12, 13
FAS, 25, 27, 30
February, 91, 92
federal funds, 62, 67, 78
federal government, 8, 9, 13, 40, 42, 73
fee, 31
feedstock, 76
fees, 31, 50, 62, 78
felony, 46
finance, 31, 45

Index

financial institution, 30
financial institutions, 30
financial resources, 32, 43, 53
financing, 30, 31
fines, 44, 45, 46
firms, 26, 28
flexibility, 3, 10, 25, 30, 36, 39, 44, 58, 72
flow, 45
fluid, 8
food, vii, 25, 28, 29, 32, 33, 34, 35, 36, 37, 38, 39, 40, 41, 42, 43, 44, 45, 46, 47, 48, 49, 50, 52, 53, 54, 55, 66, 85, 91
food aid, 25, 32, 33, 34
Food for Peace, 32
food production, 41
food safety, 29
food stamp, 35, 36, 37, 38, 39, 40, 41, 42, 43, 44, 45, 46, 47, 48, 49, 52, 53, 91
Food Stamp Act, 48, 52
food stamps, 35, 36, 38, 39, 40, 42, 43, 52, 53
foreign aid, 30
foreign policy, 30
forest management, 72
Forest Service, 66, 67, 71, 73
forestry, v, vii, 66, 71, 72, 89, 92
forests, 71, 72, 73, 74, 76
forfeiture, 2, 6, 10
fragmentation, 72
fraud, 31, 42, 52, 83
free enterprise, 33
free trade, 10
free trade agreement, 10
fruits, vii, 11, 12, 55, 85
frustration, 23
fuel, 75, 77, 78
fuel cell, 75
funding, 8, 13, 15, 16, 17, 18, 19, 21, 23, 25, 26, 27, 28, 29, 30, 33, 34, 48, 51, 54, 59, 60, 61, 62, 63, 67, 68, 71, 72, 74, 75, 76, 77, 78, 79, 80, 86

funds, 1, 21, 26, 33, 34, 43, 45, 47, 48, 55, 57, 60, 61, 66, 67, 71, 79, 84, 85

G

GAO, 70
gasoline, 77
gauge, 27
Geneva, 92
Georgia, 70
girls, 34
goals, 19, 23, 73, 81
goods and services, 73
government, viii, 2, 10, 26, 27, 28, 29, 42, 58, 62, 78
grains, 1, 2, 3, 11
grants, 33, 49, 59, 60, 61, 62, 63, 66, 67, 71, 72, 73, 75, 76, 79, 80
grassland, 18, 20
grasslands, 20, 21
grazing, 18
Ground and Surface Water Conservation Program, 16
groups, 26, 28, 29, 34, 50, 57, 84
growth, 16, 32, 72, 78, 86
GSM, 30, 31
guidance, 53

H

habitat, 16, 17, 18
handling, 30, 41
harvesting, 19, 76, 80
head, 65
health, 59, 73, 81
health care, 59
high-value products, 26
HIP, 16
Homeland Security, 68
honey, 1, 2
hospitals, 60
household, 35, 38, 43
households, 36, 37, 38, 39, 43, 44, 49

housing, 7, 58
human, 70
humanitarian, 34
hydrogen, 75

I

impairments, 22
implementation, 22, 51, 53
importer, 10
imports, 10, 28, 31, 77
imprisonment, 46
inactive, 31
incentive, 16, 21
incentives, 21, 49, 59, 76
income, 3, 5, 7, 10, 12, 19, 35, 37, 38, 39, 40, 43, 53, 80
income support, 3, 12
income tax, 37
indexing, 36, 39, 58
Indian, 51, 52, 54
industrial, 68, 81
industry, 34, 61, 76, 77, 79
inequity, 26
infancy, 19
infectious, 69
inflation, 52, 57, 58
infrastructure, 60, 61, 73, 81
innovation, 16, 21, 88
institutions, 66, 67
insurance, 5, 12, 83, 84
insurance companies, 84
integration, 5
integrity, 42, 47, 52
intelligence, 32
interaction, 12
Internal Revenue Service (IRS), 39
international, 25, 27, 28
International Trade, 28, 88
Iraq, 29
irradiation, 27
island, 68

J

Japan, 27
job training, 43
judge, 38
judges, 45
justification, 4

L

labor, 61
land, 3, 4, 7, 8, 12, 16, 18, 19, 20, 21, 23, 58, 65, 66, 67, 79, 80
landscapes, 72
language, 48
law, 1, 2, 3, 4, 10, 11, 12, 18, 32, 33, 35, 37, 40, 42, 43, 44, 45, 46, 49, 51, 52, 55, 57, 62, 69, 71, 77, 78
laws, 28, 39
LDP, 2, 11
lead, 7, 50
leadership, 65
legislation, vii, 75, 83
legislative, 55, 56, 85
lender of last resort, 57
lending, 58
likelihood, 10
limitation, 11, 18, 44
linkage, 84
livestock, 69
loan guarantees, 31, 59
loans, 1, 2, 10, 31, 33, 57, 58, 59, 60, 61, 62, 63, 78
location, 69, 70, 76
longevity, 4
long-term, 8, 33, 59
losses, 4, 6, 42, 83, 84
low-income, 48, 49, 52

M

mad cow disease, 27
maintenance, 18, 26

Index

management, 68, 81
mandates, 33, 77, 78
maritime, 34
market, 1, 2, 3, 4, 5, 8, 9, 10, 12, 22, 25, 26, 29, 30, 31, 32
market prices, 1, 2, 3, 4, 5, 10, 12
market share, 31
marketing, 2, 3, 4, 6, 9, 10, 13, 26, 30, 86
markets, 4, 22, 26, 27, 30, 49, 53, 72
measures, 16, 22, 46, 69, 83, 84
Medicaid, 38
medical care, 59
Medicare, 59
metric, 33
metropolitan area, 77
Mexican, 10
Mexico, 10
migration, 63
military, 39, 40
milk, 1, 2, 8, 9
Milk Income Loss Contract (MILC), 8, 9
models, 22
money, 9, 21, 36, 39, 48, 51, 55, 84
mouth, 69
multilateral, 28

non-profit, 26
non-tariff barriers, 27
North American Free Trade Agreement, (NAFTA), 10
nutrition, vii, 41, 44, 47, 48, 49, 50, 53, 55, 56, 85
nutrition education, 47, 48, 49
nutrition programs, 48, 50, 55, 56, 85

O

obesity, 48, 49
obligate, 41
obligations, 9, 12
Office of Management and Budget (OMB), 1, 3, 5, 6, 17, 85
online, 42
online interaction, 42
Oregon, 16
organic, vii, 86
organization, 52
organizations, 25, 28, 34, 50, 51, 72
overproduction, 2
oversight, 72, 86
ownership, 4, 57, 58, 72

N

nation, 6, 13, 18
national, 5, 46, 47, 66, 72, 77
National Security Council, 29
natural, 22, 23
natural disasters, 22, 23
NBAF, 69
negligence, 42
network, 51, 77
new, vii, 2, 3, 7, 8, 9, 14, 17, 18, 19, 20, 22, 23, 25, 26, 27, 32, 41, 42, 44, 45, 49, 50, 51, 52, 53, 54, 55, 56, 60, 61, 62, 65, 66, 67, 68, 69, 71, 72, 73, 77, 78, 81, 83, 85, 86
New York, 77
non-emergency, 33
nonfat dry milk, 8

P

packaging, 27
palm oil, 77
paper, 41
peanuts, 2, 3
penalties, 44, 47
penalty, 36, 39, 40, 47
performance, 91
periodic, 50, 53
petroleum, 68, 81
physical activity, 49
planning, 42, 61, 69, 71, 72
planting flexibility, 11, 12
plants, 62, 76, 78
play, 31

Plum Island Animal Disease Center, 68, 69
policymakers, 32
political, 5
poor, 36, 39
population, 48, 49, 70
portfolio, 15, 57
postsecondary education, 38
poultry, 55, 85
poverty, 34
power, 62, 78
preference, 69, 76
premium, 83, 84
premiums, 84
preschool, 33
pressure, 71, 84
price mechanism, 2
prices, 2, 3, 4, 5, 7, 8, 10, 11, 21, 58
priorities, 19, 28, 68, 80
private, 16, 19, 20, 22, 27, 30, 33, 34, 58, 66, 68, 72, 73, 81, 84
private sector, 22, 27
procedures, 27, 57
producers, 3, 10, 11, 12, 16, 18, 21, 26, 68
product market, 72
production, 3, 4, 5, 8, 9, 12, 19, 20, 62, 68, 70, 73, 75, 76, 77, 78, 80, 81, 83
program, vii, viii, 2, 3, 4, 5, 6, 7, 8, 9, 10, 11, 12, 13, 15, 16, 17, 18, 19, 20, 22, 23, 25, 26, 27, 29, 31, 32, 33, 34, 35, 37, 38, 39, 40, 41, 42, 44, 46, 47, 48, 49, 50, 52, 53, 54, 57, 58, 59, 60, 61, 62, 67, 71, 72, 73, 74, 76, 77, 78, 79, 80, 81, 83, 84, 86, 91
program administration, 39
promote, 42, 47, 49, 52, 55, 73
property, 7, 57, 68
protection, 13, 17, 19, 20, 77
public, vii, 22, 41, 53, 61, 66, 68, 71, 81
public investment, 22
purchasing power, 58

Q

quality control, 46
quality of life, 60

R

range, 37, 61, 77, 79
rangeland, 18, 20
Reagan Administration, 66
real property, 35, 38
real-time, 32
recognition, 41
reconstruction, 30, 59
reduction, 16, 21, 32, 47
redundancy, 32, 66
REE, 65, 89
refiners, 10
Reform Act, 65
reforms, 65, 67
regional, 6, 17, 26, 33
registries, 22
regular, 5, 37, 44, 52, 54
regulations, 17, 19, 52, 86
rehabilitate, 22
rehabilitation, 59
reimbursement, 37
renewable energy, 61, 62, 75, 76, 78, 79
Renewable Fuels Standard, 77
rent, 7
research, vii, 26, 61, 65, 66, 67, 68, 69, 71, 75, 76, 79, 80, 81, 86
research and development, 61, 75, 78, 79, 80, 89, 91
research funding, 67, 68, 76, 81
resources, 13, 17, 27, 28, 29, 30, 34, 35, 38, 66, 86
restoration, 19
retention, 12
retirement, 13, 19, 20, 35, 36, 79
revenue, 5, 6, 10, 12
RFS, 77
rice, 2, 3, 6, 11, 12

Index

risk, 5, 46, 47, 69, 71, 72
risk factors, 70
risk management, 5
risks, 5, 6, 16, 70, 73, 81
rural, vii, 58, 59, 60, 61, 62, 63, 78, 79
rural areas, 59, 61, 62, 63, 78
Rural Business-Cooperative Service, 59
rural development, vii, 58, 60, 63
Rural Housing Service, 59
Rural Utilities Service (RUS), 59

S

safety, 6, 13, 28, 59, 69
sales, 30, 53, 58
sanctions, 46, 47, 48
savings, 1, 2, 5, 7, 10, 15, 16, 32, 35, 36, 38, 39, 41, 43, 44, 47, 76
school, 33, 34, 43, 49, 53, 54, 55, 84
school meals, 43
science, 27, 66
scientific, 27, 28
scientists, 67, 68, 80, 81
scores, 21
SEA, 66
seafood, 55, 85
Secretary of Agriculture, vii, 1, 6, 14, 29, 32, 69, 75
Secretary of the Treasury, 8
security, 33, 69
seizure, 45
services, iv, 1, 5, 6, 15, 17, 20, 22, 43, 44, 50, 53, 59, 60
sharing, 4
shelter, 38, 43
shipping, 10
short-term, 3, 30, 31
sign, 2
sites, 55, 69, 84
socially, 21, 57
soil, 13, 17, 21
soil erosion, 21
solar, 62, 78
solutions, 62, 72, 77, 78
Somalia, 30
South Korea, 27
soybean, 6
soybeans, 2, 3
specialty crop, 12, 25, 26, 68
species, 18
speed, 51
spending limits, 3
spouse, 6
stability, 5
standards, 15, 22, 27, 28
state innovation, 42
statutory, 31, 57, 69
storage, 30, 76
subsidies, 3, 8, 12, 14, 77, 83, 84
subsidy, 2, 7, 9, 31, 32, 40, 83
Sudan, 30
sugar, 1, 7, 10
sugar industry, 10
supplemental, 40, 63, 83, 84
Supplemental Security Income (SSI), 35, 36, 91
suppliers, 77
supply, 10, 77
surplus, 40, 55, 56, 85
sustainability, 18
switching, 41
Switzerland, 92
systems, 21, 41, 42, 51, 59, 61, 75, 79

T

targets, 16, 63
tariff, 10, 77
tax credit, 77
taxes, 7, 37, 53
technical assistance, 16, 18, 29, 33, 48, 61, 71, 72, 73
technology, 19, 62, 71, 72, 73, 76, 77, 78, 81
telemedicine, 60
Temporary Assistance for Needy Families (TANF), 35, 36, 37, 38, 39, 43

tenants, 4
terrorism, 70
textile, 4, 11
textile industry, 4
thinking, 76
threat, 42
threatened, 6, 18, 33
threshold, 84
thresholds, 46
title, vii, 65, 75, 91, 92
Title III, v, 25, 88
trade, vii, 3, 5, 9, 12, 14, 25, 26, 28, 29, 30, 31, 32
trade agreement, 5
Trade Representative, 29
trading, 45
traffic, 45
training, 37, 47, 61
training programs, 37, 47
transfer, 12, 41
transition, 33, 41
transition countries, 33
transportation, 76
Treasury, 1
trend, 28, 42, 77
tribal, 51, 52

U

U.S. Department of Agriculture (USDA), vii, 55, 84
U.S. Treasury, 1
unemployment, 63
uniform, 22
United States, 10, 12, 14, 32, 33, 69, 77, 92
universities, 66, 67, 68, 69, 80
urban, 73
Uruguay, 14, 32
Uruguay Round, 14, 32
users, 11, 31

V

vegetables, vii, 11, 12, 49, 54, 55, 85
vehicles, 36, 38, 39
virus, 69
vouchers, 52, 53

W

Washington, 90
waste, 60, 63, 83
waste water, 60, 63
water, 13, 16, 17, 22, 61, 63
water quality, 17
watershed, 18
watersheds, 13, 18, 22
welfare, 43
wellness, 49
wellness policies, 49
wetlands, 20
wheat, 2, 3, 6, 32
WIC program, 49
wildfire, 71, 72, 73, 81
wildlife, 16, 17, 76
Wildlife Habitat Incentives Program, 16
wind, 61, 62, 78, 79
wood, 27, 71, 73, 81, 92
wool, 1, 2
workforce, 17
workload, 17
World Trade Organization (WTO), 1, 3, 9, 10, 11, 12, 14, 27, 29, 31, 32, 87
World War II, 41

Y

yield, 3, 5, 19, 79